SPECIAL PAPERS IN PALAEONTOLOGY NO. 59

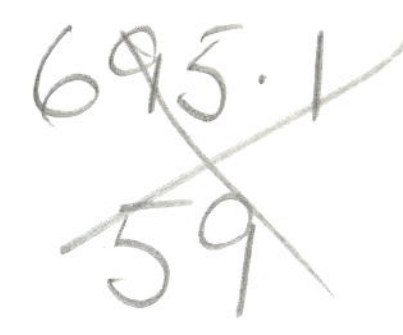

LATE ORDOVICIAN BRACHIOPODS FROM THE SOUTH CHINA PLATE AND THEIR PALAEOGEOGRAPHICAL SIGNIFICANCE

BY

ZHAN REN-BIN *and* L. R. M. COCKS

with 9 plates, 1 table and 15 text-figures

SOMERVILLE COLLEGE
LIBRARY

THE PALAEONTOLOGICAL ASSOCIATION
LONDON
JUNE 1998

© *The Palaeontological Association*, 1998

Printed in Great Britain

65772

CONTENTS

ABSTRACT. Mid Ashgill brachiopods are described from the Xiazhen, Sanqushan and Changwu formations in the border region of Zhejiang and Jiangxi provinces, East China. These are the only certain mid Ashgill shelly faunas on the South China Palaeoplate. All 41 genera recognized are represented only by single species. Strophomenoids and orthoids are the most diverse groups at 46 and 22 per cent. respectively of the taxa found, although the commonest are the rhynchonellide *Altaethyrella*, the spiriferide *Eospirifer*, and the pentameride *Tcherskidium*. Two new genera, *Rongambonites* and *Fenomena*, and seven new species are erected: *Ectenoglossa minor*, *Epitomyonia jiangshanensis*, *Rongambonites bella*, *Kassinella* (*Kassinella*) *shiyangensis*, *Sowerbyella* (*Rugosowerbyella*) *dianbianensis*, *Holtedahlina sinica* and *Fenomena distincta*. The fauna is highly endemic at the species level, but at the generic level 46 per cent. are cosmopolitan, 27 per cent. are regional and 27 per cent. are endemic. Through faunal affinity analysis of several contemporary faunas, the closest link is with the contemporary fauna from Kazakhstan.

ALTHOUGH Ordovician rocks have been known from the eastern part of Jiangxi Province and the adjacent western part of Zhejiang Province, East China (Text-fig. 1), since the pioneer work of Liu and Chao (1927), the mid Ashgill brachiopod faunas from the area were left undescribed until two species were named by Wang (*in* Wang and Jin 1964). Further work, in a palaeontological atlas of the Lower Palaeozoic of East China, specifically Jiangxi, Zhejiang, Anhui and Jiangsu provinces, included a preliminary, and in many cases sparse, account of the brachiopod fauna by Liang (*in* Liu *et al.* 1983). Since then, some individual genera and species have been described in papers by Rong *et al.* (1994), Zhan and Rong (1994, 1995) and Rong and Zhan (1996*a*). However, no complete summary of the mid Ashgill brachiopod fauna from the area has been published, and, in addition, further systematic revision has suggested reattribution and in some cases synonymy of the previously described taxa to genera known from outside China: these reasons have provided the stimulus for the present paper. Field work to review this area was undertaken by one of us (ZRB) from 1991 to 1995, guided or accompanied by Rong Jia-yu, Fu Li-pu and Zhang Yuan-Dong.

The area is important, since, during the mid Ashgill, relatively thin graptolitic shales were deposited over most of the South China Plate, and from only the present area and possibly from Henan Province (Xu 1996) are brachiopods and other shelly fossils preserved, in contrast with the latest Ashgill Hirnantian deposits, from which diverse and abundant shelly faunas are known from many parts of South China (Rong 1984*a*; Rong and Harper 1988). In the three mid Ashgill formations discussed here the shelly faunas vary greatly in both abundance and diversity, from a relatively sparse and deeper-water *Foliomena* Association (Cocks and Rong 1988) through mid-shelf associations, such as the one dominated by the pentameride *Tcherskidium*, to shallow-water nearly monospecific associations dominated by abundant forms such as the rhynchonellide *Altaethyrella*. We have taken the opportunity to contrast this South China Plate shelly fauna with other contemporary faunas recently described from North China, Kazakhstan, Taimyr, Altai and Australia, since the relationships between the various palaeoplates in what are now Australasia and east and south-east Asia have been under discussion.

GEOLOGICAL SETTING

During the mid Ashgill, the Upper Yangtze Platform to the west was extensively covered by black graptolitic shales (the Wufeng Formation) without any normal benthic shelly faunas. However, to the south-east there was land on the South China Palaeoplate, termed Cathaysia (Rong and Chen 1987; Rong *et al.* 1994). Bordering Cathaysia to the north was a shelf where the sediments bearing the brachiopods described here, together with other shelly faunas, were deposited. This shelf only existed after uplift in the late Caradoc; earlier, the same area was occupied by the Zhe-Wan (Zhejiang-Anhui) Basin with early and mid Ordovician graptolitic shales. There are three main lithological types, identified as formations (Text-fig. 2) as follows: (1) the Sanqushan Formation, of mainly carbonates; (2) the Xiazhen Formation, of intercalated clastics and carbonates; and (3) the Changwu Formation, chiefly mudstone with some siltstone intercalations. The environments of

[Special Papers in Palaeontology, No. 59 (1998), pp. 5–70, 9 pls] © The Palaeontological Association

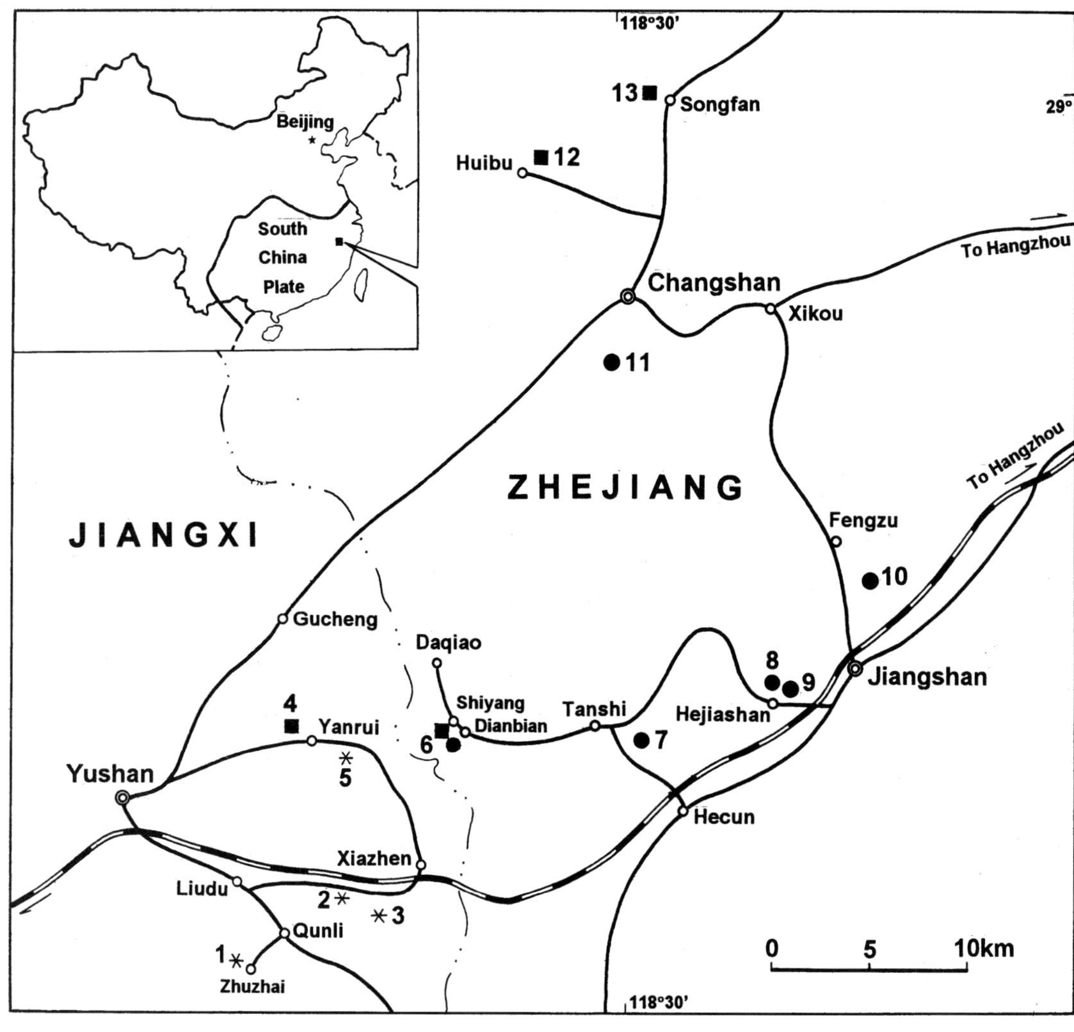

TEXT-FIG. 1. Locality map of eastern Jiangxi and western Zhejiang provinces, China, with (inset) the position of the area within the South China Palaeoplate. Numbered localities are: 1, Zhuzhai; 2, Tashan; 3, Shiyanshan; 4, Jitoushan; 5, Wangjiaba; 6, Shiyang-Dianbian; 7, Mulinlong; 8, Wujialong; 9, Pengli; 10, Shangwu; 11, Huangnitang; 12, Huishandi; and 13, Sanqushan. The dominant formation is shown for each numbered locality, except that Locality 6 is shown by two symbols since the Sanquashan Formation is intercalated within the Changwu Formation there.

the epicontinental sea were complicated, and consequently the bio- and lithofacies changed from place to place both transversely and longitudinally (Zhan and Fu 1994). All three formations are underlain conformably by the early Ashgill Huangnekang Formation and succeeded unconformably by Carboniferous or later rocks. Thirteen sections (Text-fig. 2) from this time interval at different localities were investigated and collected, and nearly 25000 specimens were obtained, of which over 90 per cent. are brachiopods. From each locality various collections were made (e.g. Collections Yz 7 and Yz 18 at Locality 1), and these are marked on each column in Text-figure 2.

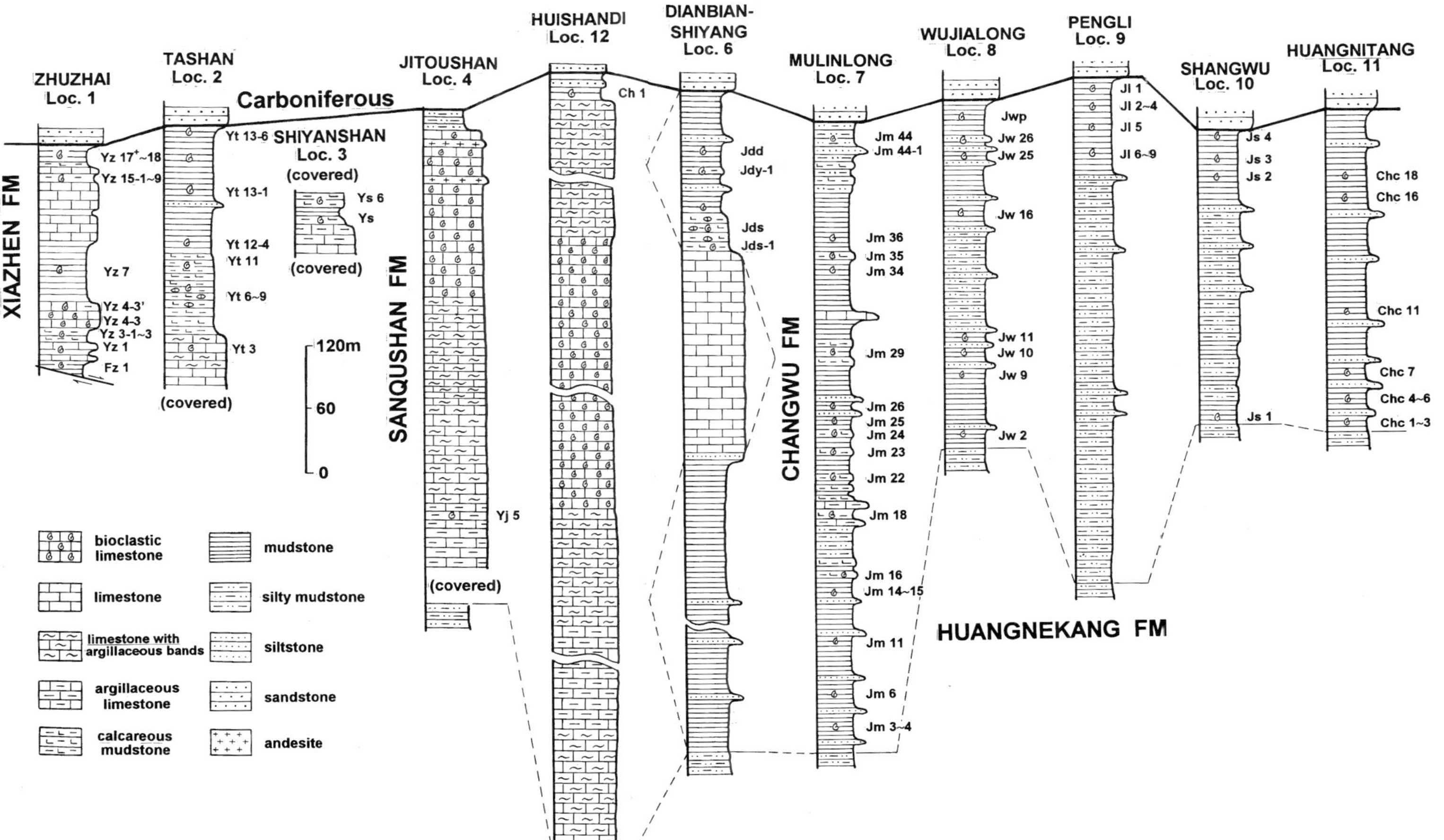

TEXT-FIG. 2. Correlation chart showing columns for Localities 1–4 and 6–12. Data are as yet incomplete for Localities 5 and 13. The geographical position of each locality is shown in Text-figure 1. Collection numbers are shown to the right of each column. The boundary between the Xiazhen and Sanqushan formations lies between Localities 3 and 4.

Locality 1. Zhuzhai of Qunli, Yushan. This is 15 km south-east of Yushan County town, where the Xiazhen Formation is 125·6 m thick as originally measured by Li and Han (1980). It was described in detail by Chen *et al.* (1987). The base is faulted against the Jurassic sandstone of the Linshan Group, and the weathered palaeocrust at its top is overlain disconformably by the Early Carboniferous Yejiantang Sandstone. Associations dominated by *Tcherskidium*, *Eospirifer* and *Sowerbyella-Antizygospira* all occur. A bioherm also occurs in the bioclastic limestone in the lower part of the section (Chen *et al.* 1987; Yu *et al.* 1992).

Locality 2. Tashan of Xiazhen, Yushan. 16 km east of Yushan County town, this is the type locality of the Xiazhen Formation and was originally measured and described by Lin and Zou (1977) and revised by Zhan and Fu (1994). This formation here is 256·5 m thick and can be subdivided into two parts: the lower part is mainly of limestone, argillaceous limestone or calcareous mudstone, whilst the upper is mostly mudstone. Its base is not exposed and its top is disconformably overlain by the Lower Carboniferous Yejiatang Formation (mainly sandstone). The shelly faunas are mainly from the argillaceous limestone, calcareous mudstone and mudstone, in which *Altaethyrella*, *Tashanomena* and *Ectenoglossa* dominate associations.

Locality 3. Shiyanshan of Xiazhen, Yushan. Shiyanshan, about 2 km east of Tashan, has an incomplete but extremely fossiliferous section through the Xiaxhen Formation, which comprises mainly bioclastic limestone in the lower part and calcareous mudstone in the middle and upper parts. Its underlying and overlying formations are not exposed. Associations dominated by *Altaethyrella* occur from the bottom to the top of this section and are especially rich in calcareous mudstone. The rocks in this section were named the Xiazhen Limestone Member by Chen (*in* Lu *et al.* 1976), upgraded to the Xiazhen Formation by Zhan and Fu (1994), who reassigned its type locality to Tashan of Xiazhen (Locality 2). Lin and Zou (1977) described corals, and Wang (*in* Wang and Jin 1964) and Liang (*in* Liu 1983) described some brachiopods including *Altaethyrella zhejiangensis*, *Sowerbyella sinensis* and *Eosotrophina uniplicata* from this section.

Locality 4. Jitoushan of Yanrui, Yushan. This section, 10 km north-east of Yushan County town, is the type locality of the Jitoushan Limestone of Chen *et al.* (1987), but that formation is included here within the Sanquashan Formation, which is more than 350 m thick and is contemporary with the Xiazhen Formation. It can be subdivided into four parts (from the bottom); light grey argillaceous limestone (18 m), dark grey limestone with rich argillaceous bands (126 m), carbonate mud mound (> 150 m) and a 55 m-thick light grey bioclastic limestone. Lai *et al.* (1993) termed the lower two parts the Yaojiaken Member and the upper two the Jitoushan Member, but we do not use these terms because they can only be recognized at this locality. Only a few brachiopods have been found in the argillaceous limestone (Collection Yj 5), including *Triplesia zhejiangenesis*, *Altaethyrella zhejiangensis*, *Eospirigerina yulangensis*, *Antizygospira liquanensis* and *Ovalospira dichotoma*. Lin and Zou (1977), Chen *et al.* (1987), Yu *et al.* (1992) and Bian *et al.* (1996) have published on the stratigraphy, mudmound and the coral fauna of this section. Two interbeds of andesite occur in the upper part of this section (Xue Yao-song, pers. comm.).

Locality 5. Wangjiaba of Yanrui, Yushan. This locality, 2 km east of Jitoushan, is not sufficiently known to provide a detailed section, but the mid Ashgill rocks of the Xiazhen Formation are very thick (over 400 m) and consist of lower light grey bioclastic and argillaceous limestones and an upper yellowish-green mudstone. The underlying formation is not exposed and the section is overlain by Early Carboniferous sandstone. The shelly fossils from the lower part are very rich both in density and diversity.

Locality 6. Shiyang and Dianbian of Daqiao, Jiangshan. Two small villages about 3 km south of Daqiao town have a series of discontinuous but fossiliferous outcrops which were treated as separate sections by Lin and Zou (1977), Qian (1987) and Lai *et al.* (1993) but are amalgamated in Text-figure 2. There is a continuous section from Shiyang to Shiyangwei, another small village 400 m west of Shiyang, which is underlain conformably by the early Ashgill Huangnekang Formation and disconformably succeeded by the Early Carboniferous Yejiatang Formation. At Dianbian there are 62 m of yellowish-green mudstone corresponding to the upper part of the Shiyang section. In the lower part, greyish-green mudstone (> 100 m) of the Changwu Formation are followed by bioclastic limestones (*c.* 150 m) of the Sanqushan Formation, which is intercalated within the Changwu Formation. At the top of the section a further 154 m of the Changwu calcareous mudstones and mudstones are seen. The brachiopods are mainly from the upper part and include associations dominated by *Altaethyrella*, *Kassinella* and *Plectorthis-Antizygospira*. The middle limestone was termed the Daqiao Limestone by Chen (*in* Lu *et al.* 1976) but included in the Xiazhen Formation by Zhan and Fu (1994) and in the Sanquashan Formation herein.

Locality 7. Mulinlong of Tanshi, Jiangshan. 2·5 km east of Tanshi town, and 9·5 km west of Jiangshan County town, is the site with the thickest sequence (*c.* 590 m) of the Changwu Formation. It is composed mainly of mudstone and calcareous mudstone with some limestone nodules, and is underlain conformably by the purple-red Huangnekang Formation and succeeded disconformably by the Early Carboniferous Yejiatang Formation. Because the strata dip at a very low angle, this section extends for nearly 2 km laterally from the bottom to the top; it is well exposed. Associations dominated by *Epitomyonia*, *Tcherskidium*, *Sowerbyella-Antizygospira* and *Eospirifer* occur.

Locality 8. Wujialong of Hejiashan, Jiangshan. 3 km west of Jiangshan County town there is a good section of the Changwu Formation (319 m thick) consisting of yellowish-green mudstone with some thin interbeds of siltstone. It is underlain conformably by the Huangnekang Formation and overlain disconformably by the Early Carboniferous Yejiatang Formation. The section is poorly fossiliferous but includes associations dominated by *Eospirifer* and *Kassinella.*

Locality 9. Pengli of Hejiashan, Jiangshan. About 700 m east of Locality 8 is a well-exposed section of the Changwu Formation at Pengli which is 473 m thick and has the same underlying and overlying rocks as at Locality 8. This section is composed mainly of yellowish-green mudstone with a few siltstone interbeds. Only the upper part is fossiliferous, from which the oldest known eospiriferine *Eospirifer praecursor* was named (Rong *et al.* 1994). *Wangyuella ventriconvexa* is the commonest brachiopod.

Locality 10. Shangwu of Fengzu, Jiangshan. 4 km north of Jiangshan County town is the type locality of the Changwu Formation (Lu *et al.* 1955) which is a yellowish-green mudstone conformably underlain and disconformably overlain by the Huangnekang Formation and the Yejiatang Formation respectively. Fossils are sparse, but Rong and Zhan (1966*b*) have identified a *Foliomena* fauna in the upper part of these beds.

Locality 11. Huangnitang of Erduqiao, Changshan. About 3 km south of Changshan County town, there is a section of the Changwu Formation at Huangnitang. Ordovician rocks are continuously exposed from the Tremadoc Yinchufu Formation to the mid Ashgill Changwu Formation which is succeeded disconformably by the Lower Carboniferous. The candidate international stratotype section for the Arenig–Llanvirn boundary is within the Ningkuo Formation here. The Changwu Formation is *c.* 150 m thick and composed of yellowish-green mudstone. It is poor in fossils, but the lower part includes *Foliomena folium*, *Cyclospira* sp., *Dedzetina* sp., *Skenidioides* sp. and others, which we identify as a *Foliomena folium* Association.

Locality 12. Huishandi of Huibu, Changshan. 11 km north-west of Changshan County town and about 7 km south-west of Songfan is the type locality of the Sanqushan Formation, transferred from Sanqushan itself by Zhan and Fu (1994). This formation is more than 1500 m thick and is composed mainly of limestone and interbedded limestone and calcareous mudstone; within the upper part there is a carbonate mudmound. It is underlain by the Huangnekang Formation, and above is succeeded conformably by 50 m of yellow mudstone assigned to the Wenchang Formation by Liang (1977) and Qian (1987). However, that formation has its type locality over 100 km to the north-east of our area and is of Hirnantian age, and we include these yellow mudstones within the Sanqushan Formation. Lai *et al.* (1993) erected new formation names at this locality, but we include them within the Sanqushan Formation.

Locality 13. Sanqushan of Songfan, Changshan. This was the original type locality of the Sanqushan Limestone (Sheng 1951), and is 13 km north of Changshan County town. It is a mountain composed of greyish-white bioclastic limestone without any underlying and overlying rocks exposed and topographically appears to be a carbonate mudmound. No brachiopods have yet been reported from it, and the section has not been measured in detail.

Correlations between these 13 sections have been published by Lin and Zou (1977), Liang (1977), Lai *et al.* (1993) and Zhan and Fu (1994). Clearly the Changwu Formation at Mulinlong (Locality 7) is the key to accurate correlation to the whole area (Text-fig. 2). The junction between the underlying Huangnekang Formation and the Changwu Formation at Mulinlong can be correlated with the base of the Sanquashan Formation at Huishandi (Locality 12), Jitoushan (Locality 4), and the base of the Changwu Formation at Dianbian-Shiyang, Wujialong, Pengli and Shangwu (Localities 6 and 8–10). The shallow-water brachiopods and corals from some layers in the middle to upper part of Locality 7 are similar to those found from the upper part of the Xiazhen Formation at Zhuzhai (Locality 1), Tashan (Locality 2) and Shiyanshan (Locality 3), and to those from the upper member of the Sanqushan Formation at Huishandi (Locality 12). Thus the middle to upper parts of the Changwu Formation at Mulinlong probably correspond to the Xiazhen Formation at Zhuzhai, Tashan and Shiyanshan, and to the upper member of the Sanqushan Formation at Huishandi.

AGE OF THE ROCKS

The age of the Sanqushan Formation and contemporary strata has been discussed by Lin and Zou (1977), Liang (1977), Chen *et al.* (1987), Rong and Chen (1987) and Lai *et al.* (1993). The underlying rocks are all purple-red mudstones of the Huangnekang Formation, which yield the brachiopod *Foliomena* fauna and the trilobite *Nankinolithus*, and are contemporary with the Lingshiang Formation of the Upper Yangtze Platform and the Tangtou Formation of the Lower Yangtze Platform, both of early Ashgill age (Zhou *in* Lu *et al.* 1976; Rong 1984*b*; Chen *et al.* 1995).

Tcherskidium, one of the most common constituents of our fauna, is reliably known from the middle Ashgill of Kazakhstan, Altai and Taimyr, the Kolymar Peninsula and Alaska (Cocks and Modzalevskaya 1997). *Foliomena* and *Kassinella* (*Kassinella*) are known to be extinct by the end of the Rawtheyan (mid Ashgill) (Rong and Zhan 1996*b*). The *Agetolites* coral fauna is widely distributed in the pre-Hirnantian rocks of Kazakhstan, Altai and Northwest China (Shaanxi and Xinjiang provinces) (Lin and Zou 1986); the associated graptolites *Normalograptus* sp., *Rectograptus socialis* (Lapworth) and *Amplexograptus* are all common representatives of the *Dicellograptus complexus* Biozone of the Wufeng Formation on the Yangtze Platform, of mid Ashgill age (Chen Xu, pers. comm.).

There are no undoubted Hirnantian rocks and fossils known from this area, nor Silurian and Devonian deposits, since it was emergent after the late Rawtheyan owing to the enlargement of Cathaysia (Rong and Cheng 1987), and most of our sections are covered disconformably by Early Carboniferous deposits. Thus a mid Ashgill age, probably late Cautleyan to mid Rawtheyan, is confirmed for all three formations within our area, although we have not been able to distinguish any finer time subdivisions.

FAUNAL ANALYSIS AND AFFINITIES

The present fauna consists of 41 brachiopod genera each represented by a single species amongst which the strophomenoids represent 46·3 per cent. and the orthoids 22·0 per cent. of the diversity, but the sole rhynchonellid (at 33·9 per cent.) and the sole spiriferid (at 22·8 per cent.) are the most abundant constituents (Text-fig. 3). Generally, beyond mid-shelf depths, the less food available, the less abundant the animals (Fürsich and Hurst 1974). The representatives of deeper-water regimes (such as *Foliomena*, *Cyclospira*, *Epitomyonia*, *Dedzetina* and *Skenidioides*) are low in number, whilst the shallower-water forms on the Zhe-Gan Platform are much more abundant and diverse and vary greatly between localities, e.g. the assemblages dominated by *Tcherskidium*, *Sowerbyella*-*Antizygospira* and *Eospirifer* at Locality 1, *Altaethyrella* at Locality 3 and *Plectorthis* at Locality 6. The most common components in our fauna are (in order of abundance) *Altaethyrella*, *Eospirifer*, *Tcherskidium*, *Antizygospira*, *Sowerbyella* (*Sowerbyella*), *Ovalospira*, *Kassinella* (*Kassinella*) and *Foliomena*.

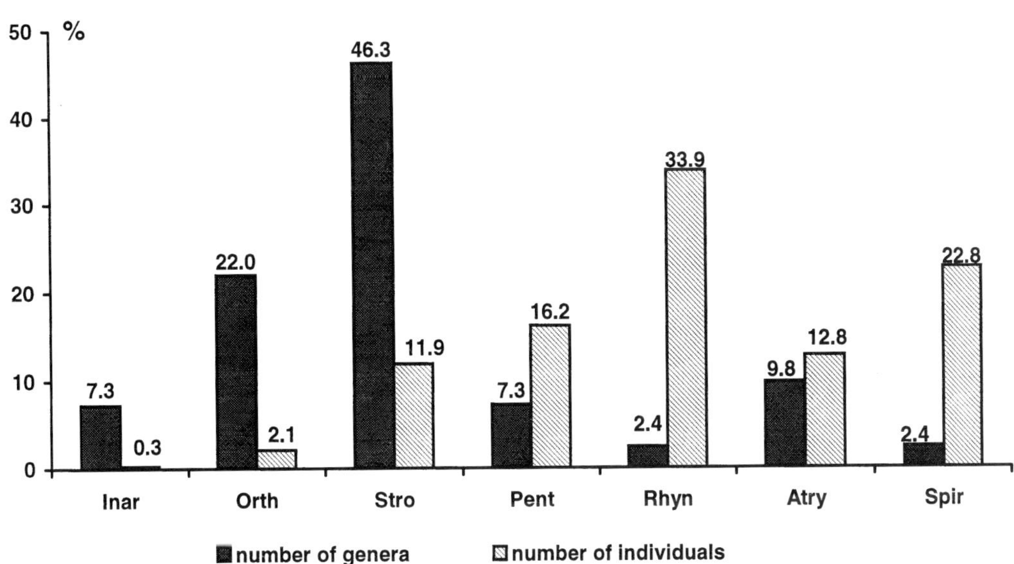

TEXT-FIG. 3. Percentages of genera and individuals for the higher brachiopod taxa in the whole of the collections from Xiazhen, Sanqushan and Changwu formations of the Zhejiang-Jiangxi border area. Inar = inarticulates, including trimerellids; Orth = Orthida; Stro = Strophomenida, including Triplesiidina; Pent = Pentamerida; Rhyn = Rhynchonellida; Atry = Atrypida; and Spir = Spiriferida.

The mid Ashgill was an interval of strong endemism and highly diversified benthic shelly faunas, particularly in epicontinental seas (Boucot 1983; Sheehan and Coorough 1990). The 41 taxa described in this paper can be divided into three groups: cosmopolitan, regional and endemic: 19 genera are largely cosmopolitan, 11 genera are found only elsewhere in palaeogeographically related faunas, and 11 taxa are endemic. All the species are endemic to the South China Palaeoplate except *Foliomena folium*, which is cosmopolitan, *Antizygospira liquanensis* and *Ovalospira dichotoma*, which also occur on the North China Palaeoplate, and *Mimella zhejiangensis*, which also occurs in Kazakhstan.

Table 1 shows the affinity indices (AI) between six middle Ashgill brachiopod faunas from the Xiazhen Formation and contemporary rocks of South China (this paper), the Beiguoshan

TABLE 1. Affinity indices between six mid Ashgill brachiopod faunas. S CHIN, the present fauna; N CHIN, Shaanxi, North China; KAZA, Dulankara, Kazakhstan; ALTA, Gorny Altai, Russia; NSW, New South Wales, Australia; TAIM, Taimyr, Siberia, Russia (for stratigraphy and references see text). Three numbers are shown for each relationship following the different formulae (below) discussed by Rong *et al.* (1995) in the lower echelon of the matrix and their averages in the upper echelon.

	S CHIN	N CHIN	KAZA	ALTA	NSW	TAIM
S CHIN	1	0·2189	0·4237	0·2777	0·0977	0·3033
N CHIN	0·1727 0·2508 0·2333	1	0·2518	0·2362	0·0250	0·0825
KAZA	0·3778 0·4559 0·4375	0·1828 0·2870 0·2857	1	0·2942	0·1935	0·2408
ALTA	0·2282 0·3063 0·2985	0·1719 0·2700 0·2667	0·2291 0·3271 0·3265	1	0·0740	0·1944
NSW	0·0519 0·1299 0·1111	-0·0511 0·0636 0·0635	0·1271 0·2313 0·2222	0·0107 0·1088 0·1026	1	0·0160
TAIM	0·2513 0·3294 0·3291	0·0305 0·1116 0·1053	0·1895 0·2706 0·2623	0·1416 0·2227 0·2188	-0·0361 0·0450 0·0392	1

Formation of Shaanxi Province, North China (Fu 1982), the upper part of the Chokpar Formation and the Dulankara Horizon at Dulankara, Kazakhstan (Nikitin *et al.* 1980; Klenina *et al.* 1984), the Orlovsk Horizon of the Altai area, Russia (Kulkov and Severgina 1989), the upper part of the Goonumbla Volcanics and the upper part of the Malongulli Formation of New South Wales, Australia (Percival 1978, 1979*a*, 1979*b*) and the Korotkinskaya Formation of central Taimyr, northern Siberia, Russia (Cocks and Modzalevskaya 1997). The lists which we have used, including some modifications from their original authors, are shown in the Appendix. Rong *et al.* (1995) compared various affinity indices and found the most useful to be Otsuka [$AI = C/(N_1N_2)^{1/2}$], Dice [$AI = 2C/(N_1+N_2)$] and Fager [$AI = C/(N_1N_2)^{1/2} - 1/(2(N_2)^{1/2})$]. In each formula N_1 is the total number of genera of one fauna, N_2 is the number of genera of another fauna and C is the number of genera common to both faunas, supposing N_2 is larger than N_1. We have calculated the affinity indices between the six faunas by all three methods and Table 1 shows each index in the lower diagonal and their average in the upper diagonal. From these results we reach the following conclusions:

1. All the indices except that between Kazakhstan and our area are less than 40 per cent, which indicates that in the late Ordovician there was strong endemism.
2. Our fauna is most closely related at the generic level to that from Kazakhstan, and is progressively less related to those from Taimyr, Altai, North China and New South Wales. However, there are only two species common to both South China and Kazakhstan, *Foliomena folium*, which is cosmopolitan, and *Mimella zhejiangensis*, which is known only from these two areas, in contrast with the three species common to both North and South China.

3. The New South Wales fauna has the lowest set of affinities to all the other five faunas.

Analysis of the corals show similar results to the brachiopods (Yang 1984), and it is reasonable to postulate that South China, Kazakhstan, North China, Altai and Taimyr had similar palaeolatitudes in the late Ordovician, and that these epicontinental benthic shelly faunas were controlled by similar environmental factors, perhaps the same ocean current.

Fortey and Cocks (1998) have reviewed aspects of the palaeobiology of south-east Asia in the Ordovician, and concluded that in the later Ordovician the palaeocontinent closest to South China was probably the Sibumasu or Shan-Thai terrane. Unfortunately, although early Ashgill *Foliomena* brachiopod assemblages and latest Ashgill *Hirnantia* fauna assemblages are known from both palaeocontinents (Cocks and Rong 1988; Cocks and Fortey 1997), there are no mid Ashgill brachiopods yet known from the Sibumasu terrane and thus direct comparison with our fauna is not possible. However, the apparent affinity distance between our fauna and those from New South Wales reinforces Fortey and Cocks's (1998) provisional conclusion that those two palaeocontinents were some distance apart in the late Ordovician. This is in contrast to the relationships proposed, for example, by Laurie and Burrett (1992), who postulated closer connections between South China and Australia, but their conclusions were supported by the relationships of only a few early Ordovician brachiopods and molluscs, rather than the more substantial faunas considered both by Fortey and Cocks (1998) and herein.

A further brachiopod fauna from the South China Palaeoplate is that recently described by Xu (1996) from the Shiyanhe Formation of Xichuan County, Henan Province, which is on the northern side of the palaeoplate. Xu has determined the age of this fauna as mid Ashgill, but this is uncertain since most of the constituent brachiopods are rather long-ranging. The most dominant forms are identified by Xu as *Nalivkinia* and *Sowerbyella*. However, that fauna (listed in the Appendix) has little similarity with the main fauna described in this paper (mean value of affinity indices 0·1972), and its significance, and indeed age, remain uncertain. The brachiopod fauna (Xu, *in* Jin *et al.* 1979) from the Koumenzi Formation (possibly mid Ashgill) of Qinghai Province, Northwest China, and situated on the Chaidam Palaeoplate, has several genera in common with our fauna: that fauna includes *Mimella*, *Triplesia*, *Sowerbyella*, *Strophomena*, *Antizygospira* (originally identified as *Zygospira* (*Sulcatospira*)) and *Eospirigerina*. But its taxonomy and age need to be revised before detailed comparisons can be made.

SYSTEMATIC PALAEONTOLOGY

All the figured specimens are deposited in Nanjing Institute of Geology and Palaeontology, Academia Sinica (NIGP), and a comparative collection is in The Natural History Museum, London (BC). All measurements are in mm, with length (L), width (W), width along hinge line where different (W_1); width of sulcus (W_2), fold (W_3) and muscle field (W_4); length/width (L/W), and thickness (T).

Superfamily LINGULOIDEA Menke, 1828
Family OBOLIDAE King, 1846
Subfamily GLOSSELLINAE Cooper, 1956

Genus ECTENOGLOSSA Sinclair, 1945

Type species. *Lingula lesueuri* Rouault, 1850, from the Arenig of France.

Diagnosis. Shell relatively long and narrow, lateral margins subparallel, ornament of concentric growth lines; two short subparallel ridges extending anteriorly from the beak of ventral valve.

Ectenoglossa minor sp. nov.

Plate 1, figures 1–3

Material. Fifteen dorsal internal and external moulds, and 21 ventral internal and external moulds; mainly from the greyish green mudstone of the upper to top part of the mid Ashgill Xiazhen Formation at Tashan of Xiazhen, Yushan (Locality 2 in Text-fig. 1, Collections Yt 13–1 ~ 6).

Description. Small to medium size, length 5–20 mm and width 2·5–7 mm: length/width ratio is 2·6–2·8 (in young individuals about 2·1). Shell outline long and tongue-like; lateral margins subparallel. Postero-lateral margins relatively short, rounded and continuous with the lateral margins at about one-seventh to one-eighth of the shell length; anterior margin arc-like. Almost equally biconvex; thickest at the shell centre with gentle lateral slopes. Ventral beak narrow; pedicle groove distinctive but short and deeply concave. Pair of thin and short ridges at both sides of the groove and subparallel to each other.

Measurements.

NIGP Cat. No.	L	W	L/W
128016 (ventral valve)	21·8	7·5	2·8
128017 (ventral valve, holotype)	18·2	6·9	2·6

Remarks. Sinclair (1945, p. 63) erected *Ectenoglossa* for lingulids with strongly extended thick shells and a pair of teeth-like structures appearing in the posterior part of the ventral interior. The new species has all these characteristics and is thus assigned to *Ectenoglossa. E. leseueuri* has been refigured from Budleigh Salterton, Devon, by Cocks and Lockley (1981) and from France by Cocks and Fortey (1988), both from rocks of Arenig age. The genus was revised by Goryansky (1969) who attributed two further species: the middle Ordovician *E. exunguis* (Eichwald) (and two Pander synonyms of that species, *angusta* and *longissima*) from the Kukruse and Idavere stages, and the lower Ordovician *E. lata* (Pander) from the Kunda and Volkhov stages, both of Estonia and north-eastern Russia. Popov (1980) described *E. sorbulakensis* from the Caradoc Anderken Stage of Kazakhstan. As far as is known, *E. minor* is the youngest representative of the genus, and the only one known from the Ashgill. It has a similar length to *E. sorbulakensis* but is relatively much narrower.

Genus PLECTOGLOSSA Cooper, 1956

Type species. Plectoglossa oklahomensis Cooper, 1956, p. 222, pl. 6c, figs 7–15; from the Bromide Formation (Caradoc) of Oklahoma, USA.

EXPLANATION OF PLATE 1

Figs 1–3. *Ectenoglossa minor* sp. nov.; Xiazhen Formation, Locality 2. 1, NIGP 128016; Horizon Yt 13–1; ventral internal mould; ×2. 2–3, NIGP 128017, holotype; Horizon Yt 13–6; internal and external moulds of ventral valve; ×2.

Fig. 4. *Plectoglossa* sp.; NIGP 128018; Xiazhen Formation, Locality 1, Horizon Yz 11; ventral internal mould; ×2.

Figs 5, 8, 11, 15–17. *Plectorthis*? *tanshiensis* (Liang, 1983); Changwu Formation, Dianbian, Locality 6, Horizon Jdd. 5, 8, 11, NIGP 128025; dorsal external and internal moulds viewed posteriorly and dorsally; ×3. 15, 17, NIGP 128026; latex cast and internal mould of dorsal interior; ×2·5. 16, NIGP 128027; ventral internal mould; ×3.

Figs 6–7. indeterminate orthine; Changwu Formation, Dianbian, Locality 6, Horizon Jdd; dorsal internal moulds. 6, NIGP 128023. 7, NIGP 128024. Both ×4.

Figs 9–10, 12–13. *Peritrimerella chunanensis* Liang, 1983; Changwu Formation. 9–10, 13, Dianbian, Locality 6, Horizon Jdd. 9, NIGP 128019; ventral internal mould; ×2; 10, 13, NIGP 128020; dorsal and posterior views of dorsal internal mould; ×2. 12, NIGP 128021; Locality 9, Horizon Jl 5; ventral internal mould of a juvenile; ×2·5.

Fig. 14. indeterminate orthoid; NIGP 128022; Changwu Formation, Dianbian, Locality 6, Horizon Jdd; dorsal external mould; ×10.

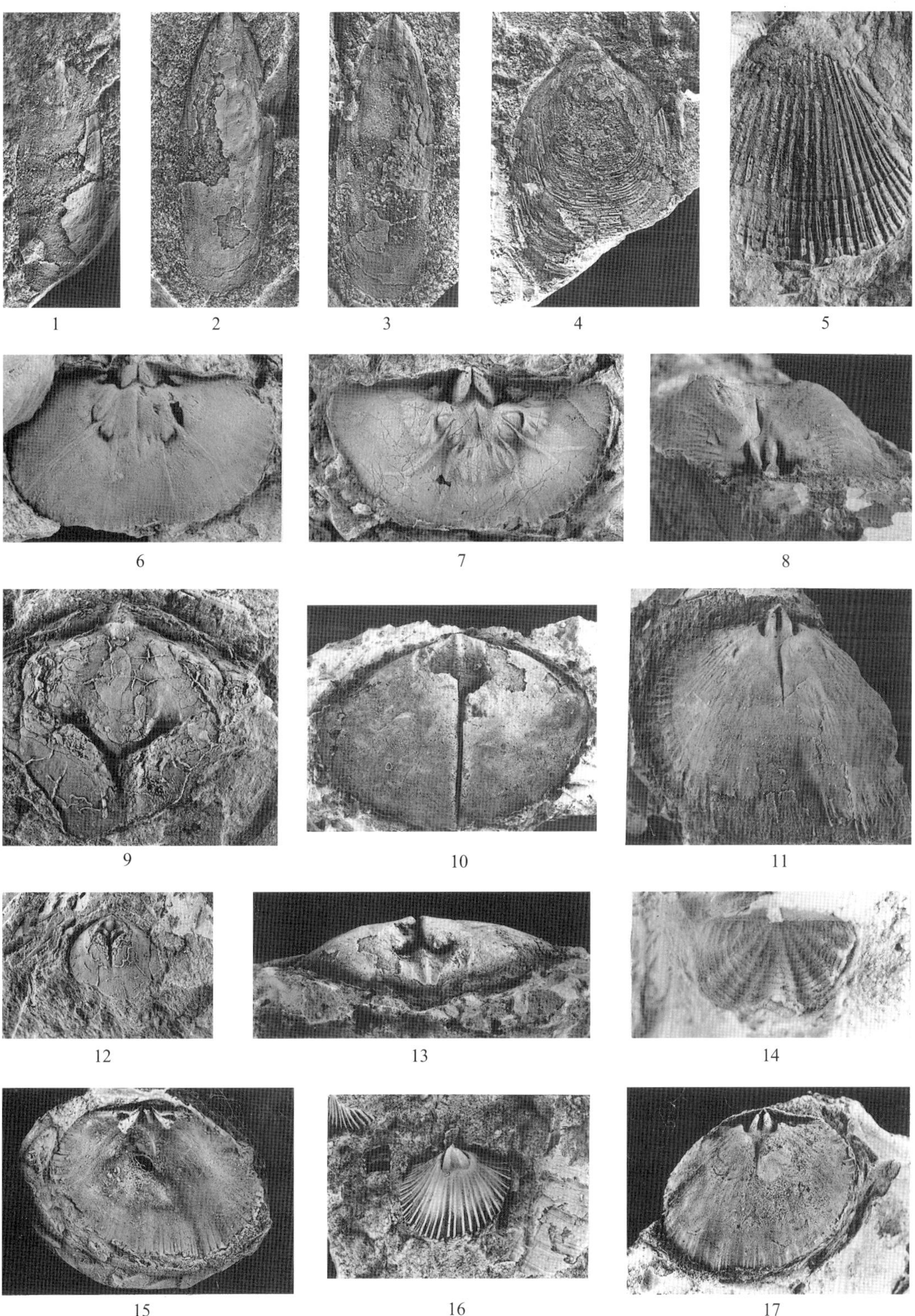

ZHAN and COCKS, late Ordovician brachiopods

Plectoglossa sp.

Plate 1, figure 4

Material. Nine specimens, from yellowish-green mudstones of the middle to upper part of the Xiazhen Formation at Zhuzhai of Qunli, Yushan (Locality 1, Collection Yz 11).

Description. Shell length 15–20 mm, width 12–16 mm; outline sub-oval with weakly differentiated postero-lateral margins; profile weakly convex. Strong concentric growth lines evenly distributed with four or five lines per mm.

Remarks. According to Cooper (1956, p. 222), *Plectoglossa* is characterized by a sub-oval outline, lens-like lateral profile and regular, elevated concentric growth lines. The present specimens have these characters and so are identified as *Plectoglossa*. Because only nine specimen have been found, many of which are not well preserved, specific identification cannot be made. The genus is known from rocks of possible Arenig to Ashgill age and is apparently cosmopolitan.

Superfamily TRIMERELLOIDEA Schuchert and LeVene, 1929
Family TRIMERELLIDAE Schuchert and LeVene, 1929
Subfamily TRIMERELLINAE Schuchert and LeVene, 1929

Genus PERITRIMERELLA Liang, *in* Liu *et al.*, 1983

Emended diagnosis. Shell small to large, outline elliptical, moderately biconvex; pseudointerarea low; anterior commissure round; ornament of few concentric fila. Ventral muscle platform large and elevated anteriorly; median ridge variably developed, but usually high and thick. Dorsal platform well developed and strongly undercut, long and high median septum supports the platform posteriorly and extends to the shell anterior margin.

Remarks. Liang's (*in* Liu *et al.* 1983) diagnosis for the type species, *P. chuanensis*, is emended here principally because our larger collections have demonstrated that a median ridge in the ventral valve is sometimes, but not always, present.

Peritrimerella chunanensis Liang, *in* Liu *et al.*, 1983

Plate 1, figures 9–10, 12–13

1983 *Petrimerella chunanensis* Liang, *in* Liu *et al.* 1983, p. 256, pl. 98, figs 10–11.

Material. Three dorsal internal moulds, 19 ventral internal and external moulds; from the yellowish green mudstone of the upper part of the Changwu Formation at Pengli of Hejiashan, Jiangshan (Locality 9, Collection Jl 5); and the upper part of the section at Locality 6, Collection Jdd, Dianbian of Daqiao, Jiangshan.

Description exterior. Most shells 2·3–15·4 mm long 3·0–22·3 mm wide; the largest known broken individual is 36·8 mm wide. Young individuals nearly round in outline (length/width 0·96) and adults transversely oval (length/width 0·70). Slightly and evenly dorsi-biconvex in lateral view; beak small and slightly curved; pseudointerareas very low (H in measurements); anterior margin round and commissure rectimarginate; shell surface with sparse concentric fila.

Ventral interior. Distinct muscle platform elevated anteriorly and slightly higher in the middle; extending to about one-third of shell length and about 35–45 per cent. of shell width. Two pairs of muscle scars, the smaller posterior and central, divided by a weak ridge which also extends anteriorly from them, and the larger lateral and slightly anterior. High and wide median ridge developed (but sometimes absent) from anteriorly of the platform to the middle of the shell.

Dorsal interior. Muscle platform well-developed, triangular, about half of shell length and one-quarter of shell width; highly elevated anteriorly with a large cavity underneath. A shallow and wide groove divides the platform into two parts. Median septum thin and high, appearing in front of the muscle platform, supporting the anterior part of the platform and extending forward to the shell anterior margin.

Measurements.

NIGP Cat. No.	L	W	L/W	H
128019 (ventral valve)	15·4	22·3	0·69	1·5
128020 (dorsal valve)	5·5	5·6	0·98	0·2
128021 (ventral valve)	?	36·8?	?	1·6

Remarks. Liang (*in* Liu *et al.* 1983, p. 256) erected *Peritrimerella* from the Changwu Formation at Chun'an, Zhejiang, East China, on the basis of large shells with elliptical outline, low and wide ventral pseudointerarea, large ventral muscle platform with a V-shaped ridge anteriorly, weak dorsal pseudointerarea, and deep and wide elevated dorsal muscle platform supported by a strong median septum anteriorly. Our specimens were collected from the same formation in the same region as those studied by Liang, but from different localities. They are very similar to Liang's specimens of *P. chunanensis*, except that they are slightly dorsi-biconvex in lateral view and, in addition, most, but not all, of the ventral moulds have a high and thick median ridge.

Superfamily ORTHOIDEA Woodward, 1852

indet. orthoid

Plate 1, figure 14

Material. Five poorly preserved dorsal external moulds; from yellowish-green mudstone in the upper part of the Changwu Formation at Dianbian of Daqiao, Jiangshan (Locality 6, Collection Jdd); and also from calcareous mudstone in the middle part of the Changwu Formation at Mulinlong of Tanshi, Jiangshan (Locality 7, Collection Jm 22).

Description. Very small, with sub-elliptical outline. Dorsal valve moderately convex, shallow dorsal sulcus starting from the umbo, and extending to the shell anterior margin where it is about 40 per cent. of the shell width. Ventral interarea large and strongly apsacline; delthyrium narrow and open. Dorsal interarea comparatively low, anacline. About ten shallow radial plicae. Within a weak sulcus a pair of plicae originate at about one-third of the shell length. Concentric accentuated growth lines evenly distributed over the shell with 13 or 14 lines per mm.

Remarks. Only exteriors are known and thus generic identification is impossible. No trace of punctae can be seen and thus the material is referred to the Orthoidea.

Family ORTHIDAE Woodward, 1852
Subfamily ORTHINAE Woodward, 1852

indet. orthine

Plate 1, figures 6–7

Material. Six dorsal valves, all internal moulds, and one external mould, from the yellowish-green mudstone of the upper part of the Changwu Formation at Dianbian of Daqiao, Jiangshan (Locality 6, Collection Jdd).

Description: exterior. Dorsal shell 6·5–8·5 mm long and 10·6–11·3 mm wide, length/width ratio 0·60–0·75. Outline semi-elliptical to semicircular. Dorsal valve flat or weakly convex, interarea large, anacline. Cardinal extremities round. Maximum width along the hingeline or slightly anterior of it. Ornament of uniform radial costae, branching once near the shell posterior part, with two per mm near the anterior margin.

Dorsal interior. Thin and high ridge-like cardinal process, with a variably developed inverted V-shaped ridge at its anterior base connecting with brachiopores laterally. Divergent brachiophores strongly projecting, with triangular bases. Sockets deep and small. Diductor muscles not only attached to the cardinal process, but also to the valve floor on both sides, occupying the whole notothyrial cavity. Adductor scars transverse, strongly impressed, sometimes slightly elevated, with variably lateral bounding ridges, about 35 per cent. of shell width; lateral and central pairs with a variably developed median myophragm. Two pairs of pallial markings digitate, branching once at their mid-length, reaching the shell anterior margin; vascula media originating from the anterior ends of the lateral pair of adductor scars, vascula myaria starting outside the middle of the inner pair of adductor scars, dividing the lateral scars and extending antero-laterally. Three genital markings saccate, well impressed, multibranching, just lateral and anterior to the muscle field; the lateral pair behind the vascula myaria.

Remarks. It is unfortunate that no ventral valves are available, because the dorsal valves show interesting pallial and genital markings which are seldom recorded. The internal structures in general are typical of the Orthinae, as revised by Jaanusson and Bassett (1993), but this form probably represents a new genus, since the only upper Ordovician genus so far identified as belonging to the subfamily is *Sulevorthis*, whose dorsal valve is more convex, its ribbing stronger, its cardinal process more robust, its pallial markings of a different pattern, and whose general valve proportions are also different from the material illustrated here.

Family PLECTORTHIDAE Schuchert and Cooper, 1931
Subfamily PLECTORTHINAE Schuchert and Cooper, 1931

Genus PLECTORTHIS Hall and Clarke, 1892

Type species. *Orthis plicatella* Hall, 1847, p. 122, pl. 32, fig. 9; from the upper Ordovician (Cincinnatian) of North America.

EXPLANATION OF PLATE 2

Figs 1–2. *Plectorthis*? *tanshiensis* (Liang, 1983); Changwu Formation, Dianbian, Locality 6, Horizon Jdd; ventral internal moulds 1, NIGP 128028; 2, NIGP 128029. Both ×3.

Figs 3–8. *Mimella zhejiangensis* (Liang, 1983); Xiazhen Formation, Locality 3, Horizon Ys. 3–4, 7, NIGP 128030; dorsal, posterior and anterior views of conjoined valves. 5–6, NIGP 128031; lateral and dorsal views of conjoined valves. 8, NIGP 128032; ventral view of conjoined valves. All ×2·5.

Figs 9–13. *Skenidioides* sp.; Changwu Formation. 9, 12, NIGP 128033; Shiyang, Locality 6, Horizon Jds; lateral and ventral views of conjoined valves; ×5. 10–11, NIGP 128034; Shiyang, Locality 6, Horizon Jds; posterior and dorsal views of conjoined valves; ×5. 13, NIGP 128035; Locality 7, Horizon Jm 1; dorsal internal mould; ×10.

Figs 14–20. *Epitomyonia jiangshanensis* sp. nov. 14–15, 18–20, Changwu Formation, Locality 8, Horizon Jw 25. 14–15, NIGP 128036, holotype; latex cast and mould of dorsal interior; ×6. 18, NIGP 128037; ventral internal mould; ×8. 19, NIGP 128038; ventral internal mould; ×6. 20, NIGP 128039; latex cast of dorsal internal mould; ×6. 16–17, NIGP 128040; Xiazhen Formation, Locality 2, Horizon Yt 12–4; external and internal moulds of dorsal valve; ×12.

Figs 21–22. *Wangyuella ventribiconvexa* Zhan and Rong, 1995; Changwu Formation, Locality 9, Horizon Jl 5. 21, NIGP 124510; posterior view of ventral internal mould; ×6. 22, NIGP 128041; ventral internal mould; ×5.

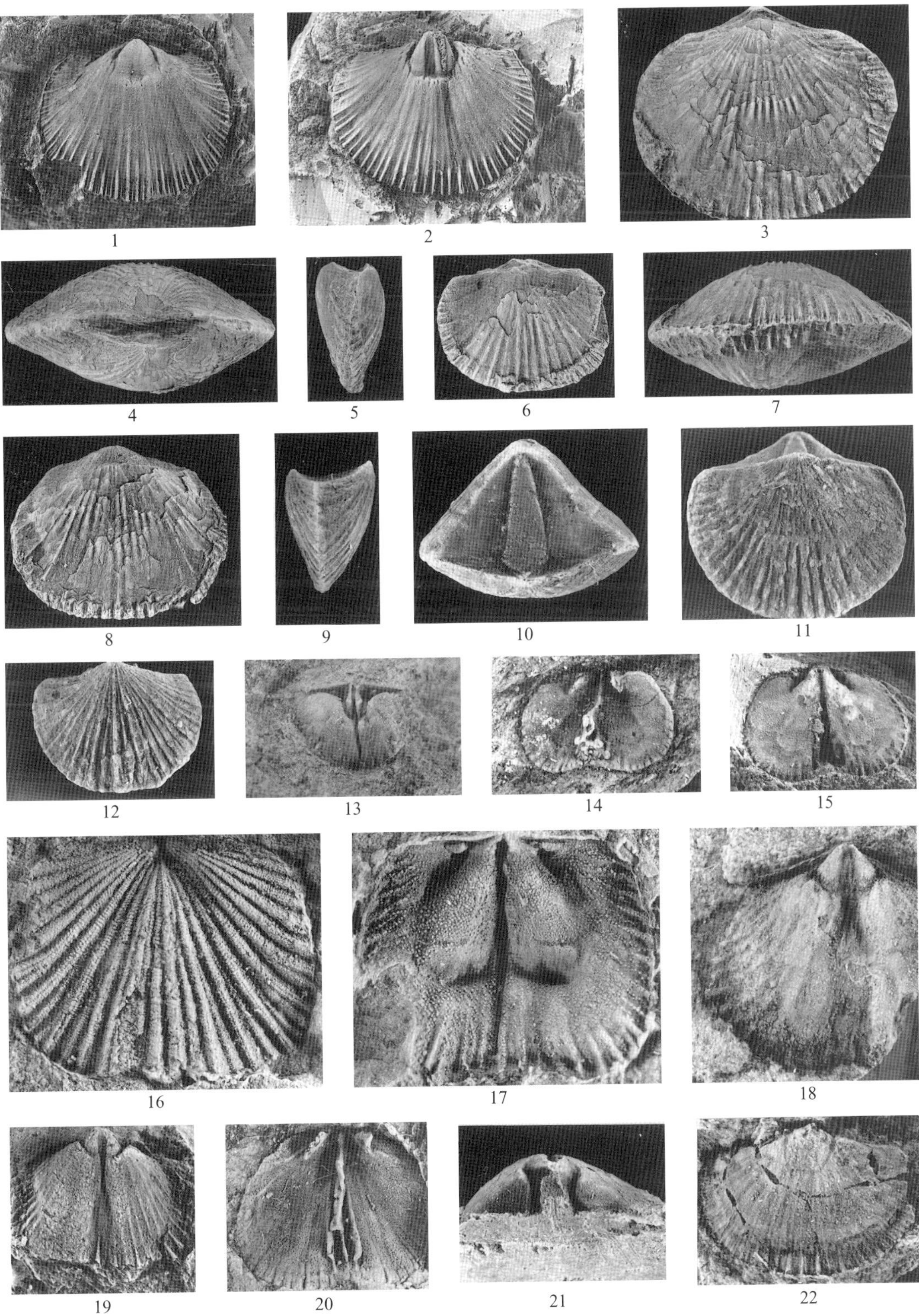

ZHAN and COCKS, late Ordovician brachiopods

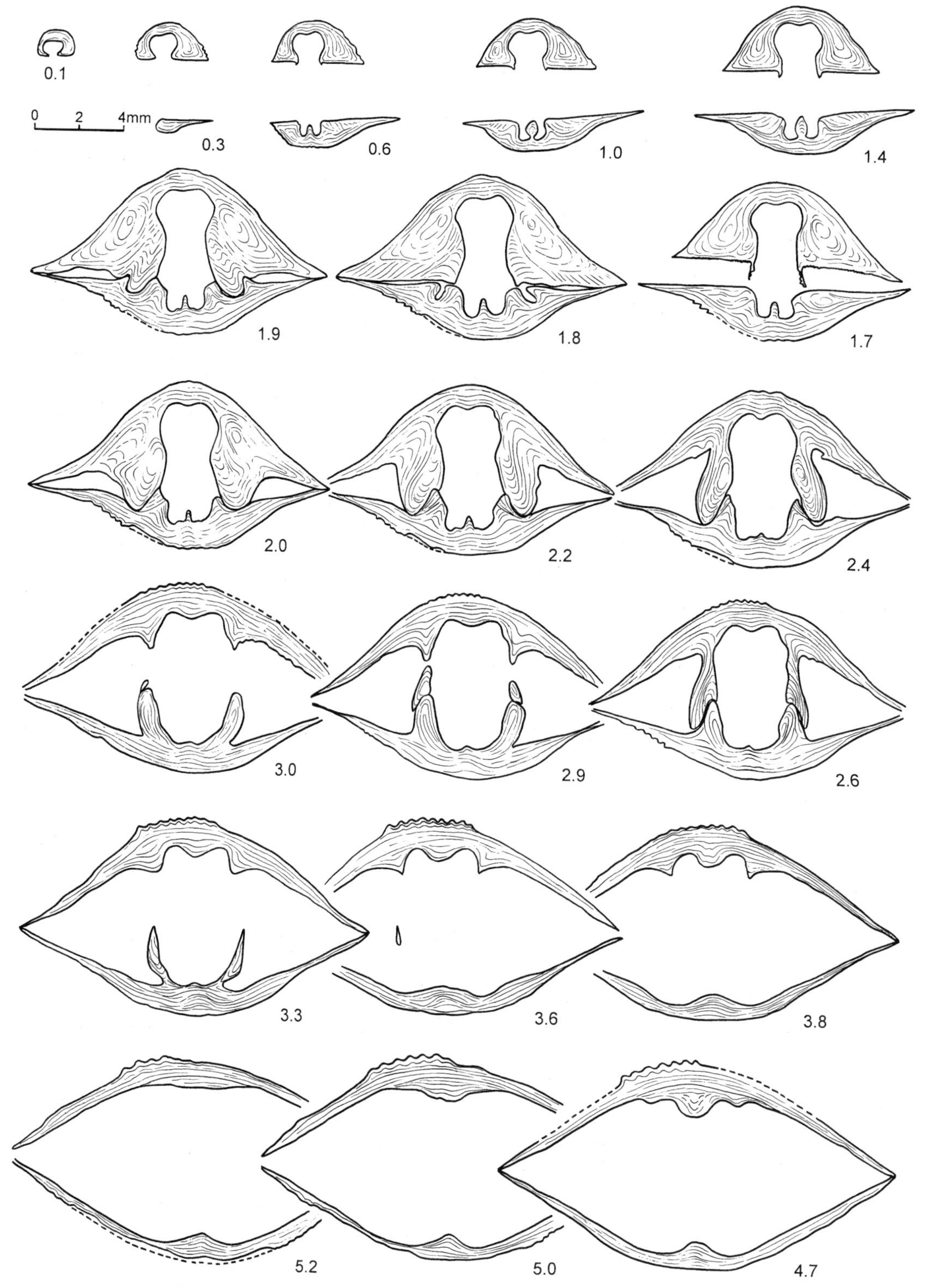
0.1
0 2 4mm
0.3
0.6
1.0
1.4
1.9
1.8
1.7
2.0
2.2
2.4
3.0
2.9
2.6
3.3
3.6
3.8
5.2
5.0
4.7

Plectorthis? *tanshiensis* (Liang, in Liu *et al.*, 1983)

Plate 1, figures 5, 8, 11, 15–17; Plate 2, figures 1–2

1983 *Plactorthis* [*sic*] *tanshiensis* Liang, *in* Liu *et al.*, p. 267, pl. 98, figs 6–7.

Material. Forty-five dorsal internal and 34 external moulds, 33 ventral internal and 21 external moulds; mainly from the yellowish-green mudstones of the upper part of Changwu Formation at Dianbian of Daqiao, Jiangshan (Locality 6, Collection Jdd). The type locality of this species is Mulinlong of Tanshi, Jiangshan (Locality 7).

Description: exterior. Shell 8–13 mm long and 10–16 mm wide; the largest shell is 15 mm long, 18 mm wide and 9 mm wide along the hinge line. The length/width ratio is 0·71–0·85. Strongly biconvex in lateral profile. Both dorsal and ventral interareas are relatively large; the ventral one is curved and anacline with a long and narrow delthyrium. Cardinal extremities round; the maximum width near the mid-length. Ventral valve evenly convex with a faint anterior fold; dorsal valve generally has a shallow and variable sulcus usually originating from mid-valve length. Ornament of radial hollow costae with smaller costellae; generally branching twice at approximately one-third and two-thirds of the shell length respectively, with a few costae branching only once and a small number three times. Generally 9–11 costae per 3 mm, 5 mm away from the umbo. Concentric fila extremely fine, evenly distributed and 10 or 11 per mm longitudinally.

Ventral interior. Teeth stout, dental plates well-developed and coalescing anteriorly with the muscle field bounding ridges. Muscle field about one-quarter to one-fifth of the valve length; the adductor scars strongly elevated and distinguished from the lateral diductor scars by a pair of weakly developed low lateral ridges; the anterior surrounding ridges of the diductor scars variable in strength. Pallial markings usually not preserved.

Dorsal interior. Cardinal process thin, ridge-like with some small crenulations on both sides of the myophore, occupying the whole notothyrial platform. Sockets relatively small, limited by fulcral plates. Subparallel brachiophores stout, extending forward, their supports high and strong and extending to *c.* 15–20 per cent. of shell length. Cardinal process connected with a low and wide myophragm which becomes a thin and high septum in the middle of the muscle scar to merge with the valve floor anteriorly. Adductor scars poorly developed, the posterior pair relatively larger. Pallial markings originating from both sides of the notothyrial platform and the anterior margin of the adductor scars, and branching several times.

Measurements.

NIGP Cat. No.	L	W	L/W	W_1
128025 (dorsal valve)	12·8	18·0	0·71	9·2
128026 (dorsal valve)	10·6	12·7	0·83	6·6
128027 (ventral valve)	5·2	6·1	0·85	4·1
128028 (ventral valve)	8·6	10·2	0·84	6·1
128029 (ventral valve)	9·0	10·8	0·83	7·7

Remarks. Most species of *Plectorthis* have subparallel brachiophore supports (Cooper 1956; Wright 1964; Mitchell 1977; Percival 1991). The present species is similar to *P. plicatella* (Hall) in its teeth, dental plates, sockets, fulcral plates, muscle scars and myophragm; however, it differs in the flaring direction and the development of the brachiophore supports. Thus we assign it provisionally to *Plectorthis*.

The type locality of *P.*? *tanshiensis* is at Mulinlong (Locality 7), 15 km east of Dianbian (Locality 6) where our material was collected; the type specimens are much larger than the present specimens. The most important characters of this species are its strong convexity, the ridge-like cardinal process which occupies the whole notothyrial cavity, the subparallel brachiophore supports, and the multibranching radial costae with additional costellae. The type species, *O. plicatella*, can be

TEXT-FIG. 4. For caption see opposite.

TEXT-FIG. 4. Transverse serial sections of *Mimella zhejiangensis* (Liang, *in* Liu *et al.*, 1983), with ventral valve upwards (36 sections made and 20 selected herein): from the Xiazhen Formation, Locality 3, Collection Ys.

distinguished from *P.*? *tanshiensis* by having costae with fewer branches, and the brachiophore supports extending medially to meet at the median ridge as do those of *P.*? *perditosulcata* and *P. cliefdenensis*. Besides the type species, 14 species of *Plectorthis* were recognized by Cooper (1956, pp. 447–456), and other species of the genus include *P.*? *perditosulcata* Wright (1964, pp. 194–196, pl. 6, figs 2–6, 8–9), *P. cliefdenensis* Percival (1991, pp. 124–126, fig. 8 (1–16)), and *P.* cf. *scotica* (McCoy) (*see* Mitchell 1977, p. 42, pl. 6, figs 27–29). All of these are less convex than our new species, and some of them have unbranching coarse plications; some have brachiophore supports which combine to form a transverse ridge subparallel to the hinge line, and some have marked dorsal transverse muscle ridges. None of them has the distinctive ornament of *P.*? *tanshiensis*. This is one of the stratigraphically highest species within *Plectorthis* and its close allies.

Genus MIMELLA Cooper, 1930

Mimella zhejiangensis (Liang, in Liu *et al.*, 1983)

Plate 2, figures 3–8; Text-figure 4

1983 *Zhejiangorthis zhejiangensis* Liang, *in* Liu *et al.*, p. 269, pl. 99, figs 4–10.

Material. One hundred and two individuals with conjoined valves and 29 external and internal moulds; mainly from the greyish-green calcareous mudstone of the lower to middle part of the Xiazhen Formation at Shiyanshan of Xiazhen, Yushan (Locality 3, Collection Ys). Liang's type locality is Locality 7.

Description: exterior. Shells usually 12–14 mm long, 14–18 mm wide and 6·5–8·5 mm thick, but the largest is 15·4 mm long, 20·2 mm wide and 8·6 mm thick and the smallest 5·3 mm long, 6·7 mm wide and 3·4 mm thick. The length/width ratio is 0·75–0·85. Transversely elliptical or nearly rectangular outline and dorsi-biconvex in lateral profile. Weak ventral sulcus and dorsal fold. Ventral interarea apsacline (height H in measurements) and dorsal interarea anacline. Delthyrium and notothyrium open with small delthyrial ridges. Ornament of radial costae, often branching twice; 11–15 in number per 5 mm, 10 mm from the umbo. The costae are variable in detail, reflecting the hollow ribs characteristic of many plectorthids.

Ventral interior. Teeth strong, dental plates thick and extending forward subparallel to form the lateral ridges of the muscle field. Muscle scars distinct, about one-third of the shell length, with no myophragm. Adductor scars become more elevated anteriorly. From five casts obtained by processing with hydrochloric acid and from the serial sections (Text-fig. 4), it appears that the width of the adductor scars has some variation: some individuals become wider from back to front, others only change slightly anteriorly.

Dorsal interior. Cardinal process well-developed; shaft and myophore distinct with the latter enlarged to form a knob at its top. The base of the cardinal process extends forward to occupy the whole notothyrial platform without any substantial change in its size. Notothyrial platform formed by the brachiophore supports extending medially along the shell floor and combining centrally. Brachiophores widely divergent. Low and thick median ridge in front of the notothyrial platform, becoming thinner and higher anteriorly to form a distinct myophragm. Two pairs of adductor scars poorly impressed and without any antero-lateral surrounding ridges; the anterior pair slightly smaller than the posterior pair. The whole muscle field is less than one-third of the shell length and width.

Measurements.

NIGP Cat. No.	L	W	L/W	T	W_1	H
128030 (conjoined valves)	13·9	17·9	0·78	8·2	11·2	1·7
128031 (conjoined valves)	8·4	11·3	0·74	4·7	8·1	1·8
128032 (conjoined valves)	10·8	14·2	0·76	6·4	9·0	1·5

Remarks. Klenina (*in* Klenina *et al.* 1984) described two species of *Mimella* from the upper Ordovician of Kazakhstan: *M. brevis* Rukavishnikova and a new species, *M. recta*. *M. brevis* is distinct from *M. recta* in the stronger ventral sulcus and dorsal fold, and the denser costae which number 26 per 5 mm, 10 mm from the umbo. From the middle and upper Ordovician of the Siberian Platform, Andreeva (*in* Nikiforova and Andreeva 1961) reported four species of *Mimella*, of which *M. panna* from the Llandeilo–Caradoc Chertov Horizon has a very strong sulcus and fold and very fine and numerous costae. Andreeva erected *sibirica* as a subspecies of *gibbosa* Billings; however, since *gibbosa* is a dalmanelloid (Hall and Clarke 1892), Billings' name is inappropriate for the Siberian form from the Ashgill Dolbor Horizon, which we thus term *Mimella sibirica*.

Cooper (1956) recognized 28 species of *Mimella* amongst which 25 species were named and three left in open nomenclature. All of these species have the distinct cardinalia, ventral muscle scars and pallial markings characteristic of *Mimella*, but they also have denser costae, a much higher ventral interarea and a more convex shell outline than the Chinese form considered here. *Mimella extensa* Cooper was also identified in China by Xu (*in* Jin *et al.* 1979, p. 72, pl. 8, figs 25–29; pl. 19, figs 1–4) from the upper Ordovician of Qilian of Qinghai: this form is larger and more transverse in outline, has a larger sulcus and fold, and has much wider ventral muscle scars with a larger bounding ridge independent of the brachiophore supports. The only other species of *Mimella* recorded from China was *M. formosa* from the Arenig Dawan Formation (Wang 1956); however, this species was later and correctly assigned to the new genus *Pseudomimella* by Xu and Liu (1984).

The material from Jiangxi Province is the same as that described by Liang (in Liu *et al.* 1983) from Zhejiang Province as *Zhejiangorthis zhejiangensis*. We do not consider that Liang's genus is substantially different from *Mimella* and use the latter name here. The species is very close to *M. recta* and the two may be synonyms, although we consider that the Kazakhstan material is of lesser convexity and might be identified as the subspecies *M. zhejiangensis recta*. *M. zhejiangensis* differs from *M. sibirica* in having a weaker fold and sulcus, and in having more costa (11–15 per 5 mm in contrast with 6–10 in *M. sibirica*). Xu (1966, p. 552, pl. 2, figs 14–31) also identified some specimens as *Mimella* cf. *recta* from the upper member of the Shiyanhe Formation (middle Ashgill) of Xichuan, Henan Province, which differ from *zhejiangensis* in having denser and more branching costae.

Family SKENIDIIDAE Kozłowski, 1929

Genus SKENIDIOIDES Schuchert and Cooper, 1931

Type species. *Skenidioides billingsi* Schuchert and Cooper, 1931; from the Black River Formation (Caradoc), near the Ottawa River, Quebec, Canada.

Skenidioides sp.

Plate 2, figures 9–13

Material. Four specimens with conjoined valves, one ventral valve, three ventral internal and external moulds respectively, and three dorsal internal and external moulds respectively; mainly from the yellowish green mudstone or calcareous mudstone of the middle to upper part of the Changwu Formation at Dianbian-Shiyang of Daqiao, Jiangshan (Locality 6, Collections Jds and Jdd).

Remarks. These specimens are the first *Skenidioides* reported from the Ashgill of China. *Skenidioides* is a well-known and cosmopolitan genus which ranges from the lower Ordovician to the upper Silurian; there are 13 middle Ordovician species described by Cooper (1956) from North America and ten named species and three unnamed species from Britain listed by Cocks (1978). Our new material represents a species with rounded cardinal extremities, almost no fold and sulcus and a large curved erect ventral interarea. It may represent a new species, although there appears to be considerable intraspecific variation in the published illustrations of the genus; however, it is not very morphologically distant from the nearly contemporary fossils identified as *S. asteroideus* (Reed) from the Ashgill of Scotland (Harper 1984) and Ireland (Wright 1964). The type species, *S. billingsi*, can be distinguished from our material by its comparatively coarser radial costae which bifurcate only once and a shallow and narrow fold originating from the middle of the ventral valve.

Superfamily DALMANELLOIDEA Schuchert, 1913
Family DICOELOSIIDAE Cloud, 1948

Genus EPITOMYONIA Wright, 1968

Type species. Epitomyonia glypha Wright, 1968 from the Boda Limestone (Ashgill) of Dalarna, Sweden.

Epitomyonia jiangshanensis sp. nov.

Plate 2, figures 14–20

Material. Sixteen ventral internal and 12 external moulds, 13 dorsal internal and eight external moulds; mainly from the yellowish-green mudstone of the middle to upper part of the Changwu Formation at Mulinlong of Tanshi and Wujialong of Hejiashan, Jiangshan (Localities 7 and 8, Collections Jm 1, Jm 6 and Jw 25).

Description: exterior. Shell 3·1–4·3 mm long and 4·0–4·8 mm wide. Sub-circular outline; ventri-biconvex in lateral profile: dorsal valve slightly convex with a very weak concave sulcus anteriorly; ventral valve strongly convex. Median sulcus relatively shallow for the genus. Hingeline about 60 per cent. of maximum width, which is usually at about 70 per cent. of valve length. Cardinal extremities rounded. Ventral interarea high, apsacline; dorsal interarea narrow, anacline. Anterior commissure slightly bilobed. Ornament of radial costae which are fine and evenly distributed; the costae bifurcate at about two-thirds of shell length. Fine concentric growth lines present. Densely populated punctae over the whole shell interior except for the muscle field.

Ventral interior. Teeth triangular, dental plates small and forming the outer ridges of the muscle field anteriorly. Diductor scars strongly impressed. Adductor scars very small but slightly longer than the larger diductor scars. Thin and high median septum originating near the umbo becoming a low and wide median ridge anteriorly which continues to the anterior margin and has a very shallow and narrow groove.

Dorsal interior. Small cardinal process connected to the median septum anteriorly. Large triangular brachiophores with their supports extending to about one-quarter to one-fifth of the shell length: the angle between them is about 90°. Sockets small, outside the brachiophores and surrounded by low ridges connected to the brachiophores. Median septum well developed, starting in front of the notothyrial platform and becoming thinner and higher rapidly anteriorly, sometimes branching anteriorly. Side septa variably developed, sometimes absent. Adductor scars very small, limited to the notothyrial cavity, with anterior surrounding ridges variably developed from strong to nearly absent.

Measurements.

NIGP Cat. No.	L	W	L/W	W_1
128036 (dorsal valve, holotype)	3·1	4·4	0·70	3·0
128037 (ventral valve)	4·3	5·1	0·84	2·7
128038 (ventral valve)	3·9	4·8	0·81	?
128039 (dorsal valve)	4·4	5·3	0·83	2·9
128040 (dorsal valve)	3·4	4·0	0·85	2·5

Remarks. Since Wright (1968) named *Epitomyonia* from the Ashgill of Sweden, this genus has been reported from many Ashgill to Wenlock localities internationally. However, this is the first record from South China. Zhang and Boucot (1988) reviewed the genus and found that the Silurian species had three distinct forms of dorsal interiors with a variety of dorsal ridges (comparable to the side septa of some Plectambonitoidea; see Cocks and Rong 1989). However, they thought that all the upper Ordovician species could be included within the form lacking dorsal ridges. Our new species is the first to be recorded from the Ordovician in which dorsal side septa are present in some specimens, although the variation is such that these septa are both present and absent in different specimens from the same locality.

The named Ordovician species of *Epitomyonia* are the type species *E. glypha* from Dalarna, Sweden (Wright 1968), *E. reclina* from the Klamath Mountains, California, USA (Potter 1990), *E. dorsicava* from the Králův Dvůr Formation, Czech Republic (Havlíček and Mergl 1982), and *E. americana* from the White Head Formation of Percé, Canada (Sheehan and Lespérance 1979). All are of Ashgill age. Our new material from China differs from *E. glypha* in the variable presence of dorsal side septa, in the finer and more numerous costae and in a less concave dorsal valve. From *E. reclina* it differs in the less indented anterior commissure, the wider hinge line and the general outline; in fact we wonder whether *E. reclina* might be better assigned to *Dicoelosia*. From *E. dorsicava* it again differs in the variable presence of dorsal side septa and has a more pronounced ventral septum, although the general shape and ribbing are very similar in both species. From *E. americana* it differs in having much stronger ventral and dorsal median septa, in the variable presence of dorsal side septa, and in having a more rounded shell outline.

Amongst the many Silurian species, our material is similar to *E. pachytriseptata* Zhang and Boucot (1988, pp. 753–758, fig. 1·9–1·15) from the late Llandovery Cape Phillips Formation of Arctic Canada, and *E. triseptata* Lenz (1977) from the Middle Llandovery of the Canadian Cordillera, in the presence of side septa, but these two species have many tubercles in their dorsal interiors. In addition, the side septa of *E. pachytriseptata* are almost as high as the median septum and *E. triseptata* has an additional pair of side septa outside the central side septa. Thus we establish here a new species, *E. jiangshanensis*, from the upper Ordovician of South China.

Family CHRUSTENOPORIDAE Baarli, 1988

Genus WANGYUELLA Zhan and Rong, 1995

Wangyuella ventribiconvexa Zhan and Rong, 1995

Plate 2, figures 21–22; Plate 3, figures 1–3

1995 *Wangyuella ventribiconvexa* Zhan and Rong, p. 557, pl. 1, figs 1–7, 10–11, 14–15; text-figs 3–5.

Remarks. Zhan and Rong (1995) erected the monospecific genus *Wangyuella* from the present fauna, with its type horizon and locality from the Changwu Formation at Pengli, west of Jiangshan County Town (Locality 9, Collection Jl 5). As far as is known, both genus and species are endemic to South China. One feature not mentioned in the original description is the variation in development of the cardinalia. The cardinalia of *W. ventribiconvexa* are generally one-sixth to one-seventh of the shell length, and some are even shorter, about one-eighth of the shell length. But there are some adult individuals from the same locality and horizon as the type which have rather large cardinalia, about one-quarter to one-fifth of the shell length (see Pl 3, fig. 1). The ?*Lordorthis* sp., of Xu (1996, p. 552, pl. 1, figs 10–12), from the upper member of the Shiyanhe Formation (middle Ashgill) of Xichuan, Henan Province, is very similar to *W. ventribiconvexa* in shell outline, cardinalia and ventral muscle field, but differs in its coarser and less branching ribs.

Family RESSERELLIDAE Walmsley and Boucot, 1971

Genus DEDZETINA Havlíček, 1951

Dedzetina? sp.

Plate 3, figure 4

Material. A single ventral valve, both internal and external moulds, from the yellowish-green mudstone of the lower part of the Changwu Formation at Huangnitang of Erduqiao, Changshan (Locality 11, Collection Chc 13).

Description. Shell 2·6 mm long and 2·9 mm wide. Ventral valve strongly convex. Cardinal extremities round. Ornament of quite coarse costae, branching only once near the umbo. Teeth strong, without dental plates. Muscle field well impressed with weak surrounding ridges; adductor scars slightly higher than the lateral pair of diductor scars.

Remarks. The present specimen resembles in some aspects *Dedzetina* as illustrated by Havlíček (1977, pls 28–29). However, without a dorsal valve, firm generic attribution and specific identification are impossible.

indet. dalmanelloid

Plate 3, figure 5

Material. Two dorsal internal moulds, from the yellowish-green mudstone of the upper part of the Changwu Formation, Dianbian of Daqiao, Jiangshan (Locality 6, Collection Jdd).

Description. Shell *c*. 3·0 mm long and 4·2 mm wide; maximum width slightly anterior to the hingeline. Dorsal interarea curved, comparatively high. Cardinal extremities round. Cardinalia 27–30 per cent. of shell length and 16–20 per cent. of shell width. Cardinal process strong and limited to the notothyrial cavity; highly projecting myophore with an extremely narrow groove in the middle which makes it bilobed. Brachiophores stout, with coarse and long processes; their supports thick and high, extending parallel anteriorly and ending sharply before the muscle field. Adductor scars very long, kidney-like with a shallow and wide median ridge in between, terminating at 90 per cent. of the shell length.

Remarks. This material is different from the other dalmanelloids described above. No ventral valves or exteriors have been recovered and one of the two dorsal interiors is illustrated here for comparison. The long parallel brachiophore supports are reminiscent of those of some draboviids.

EXPLANATION OF PLATE 3

Figs 1–3. *Wangyuella ventribiconvexa* Zhan and Rong, 1995; Changwu Formation, Locality 9, Horizon Jl 5. 1–2, NIGP 124505; internal and external moulds of dorsal valve; ×5. 3, NIGP 124504; dorsal internal mould; ×8.

Fig. 4. *Dedzetina*? sp.; NIGP 128042; Changwu Formation, Locality 11, Horizon Chc 13; ventral internal mould; ×13.

Fig. 5. indeterminate dalmanelloid; NIGP 128043; Changwu Formation, Dianbian, Locality 6, Horizon Jdd; dorsal internal mould; ×10.

Figs 6–11. *Triplesia zhejiangensis* Liang, 1983; Changwu Formation, Locality 6. 6, NIGP 128044; Horizon Jdd; ventral internal mould; ×6. 7, NIGP 128045; Horizon Jds; ventral view of conjoined valves; ×2·5. 8–11, NIGP 128046; Horizon Jds; anterior, dorsal, posterior and lateral views of conjoined valves; ×2·5.

Fig. 12. *Oxoplecia* sp.; NIGP 128047; Xiazhen Formation, Locality 1, Horizon Yz 18; dorsal internal mould; ×3.

Fig. 13. *Bimuria*? sp.; NIGP 128048; Changwu Formation, Dianbian, Locality 6, Horizon Jdd; dorsal internal mould; ×2·5.

Figs 14–17. *Metambonites meritus* Rong and Zhan, 1995; Changwu Formation, Dianbian, Locality 6, Horizon Jdd. 14, NIGP 124521; dorsal internal mould; ×3. 15, NIGP 124523, holotype; latex cast of dorsal internal mould; ×5. 16, NIGP 124531; ventral internal mould; ×4. 17, NIGP 124512; latex cast of ventral internal mould; ×5.

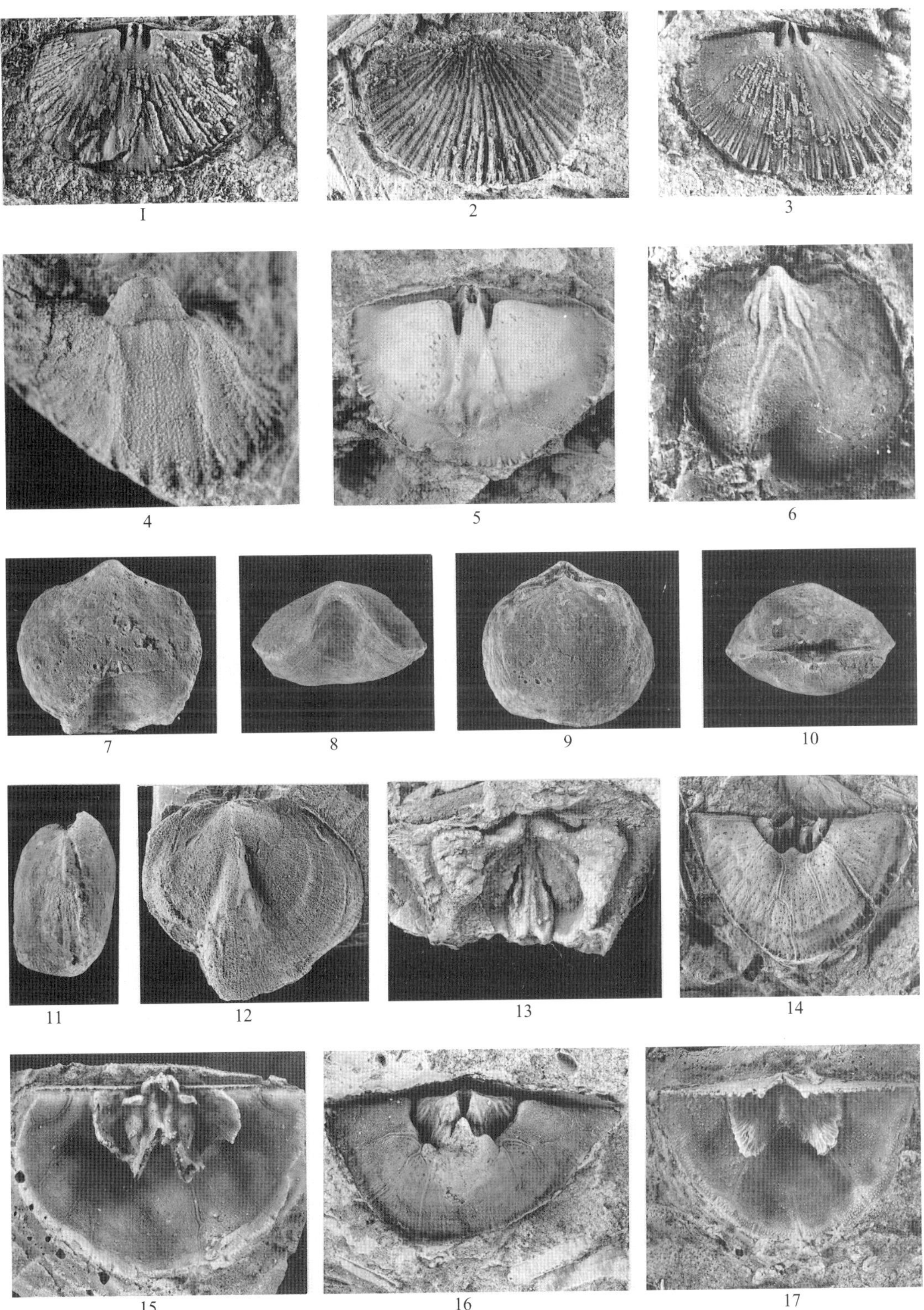

ZHAN and COCKS, late Ordovician brachiopods

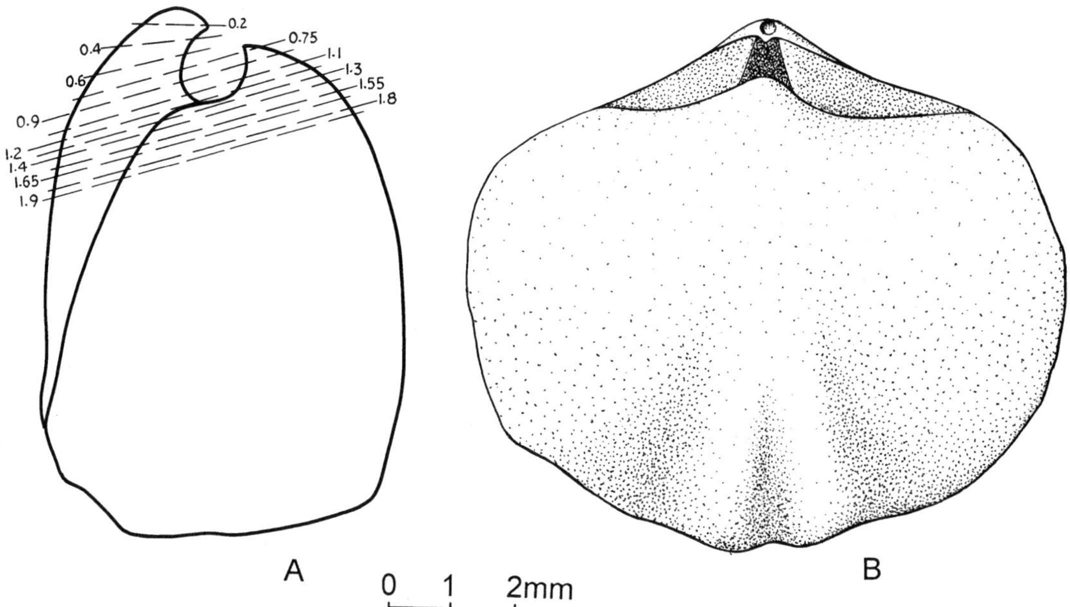

TEXT-FIG. 5. Lateral and dorsal views of conjoined valves of *Triplesia zhejiangensis* Liang, *in* Liu *et al.*, 1983, with positions of the transverse serial sections of Text-figure 6 shown on the lateral view; from the Changwu Formation, Locality 6, Collection Jds.

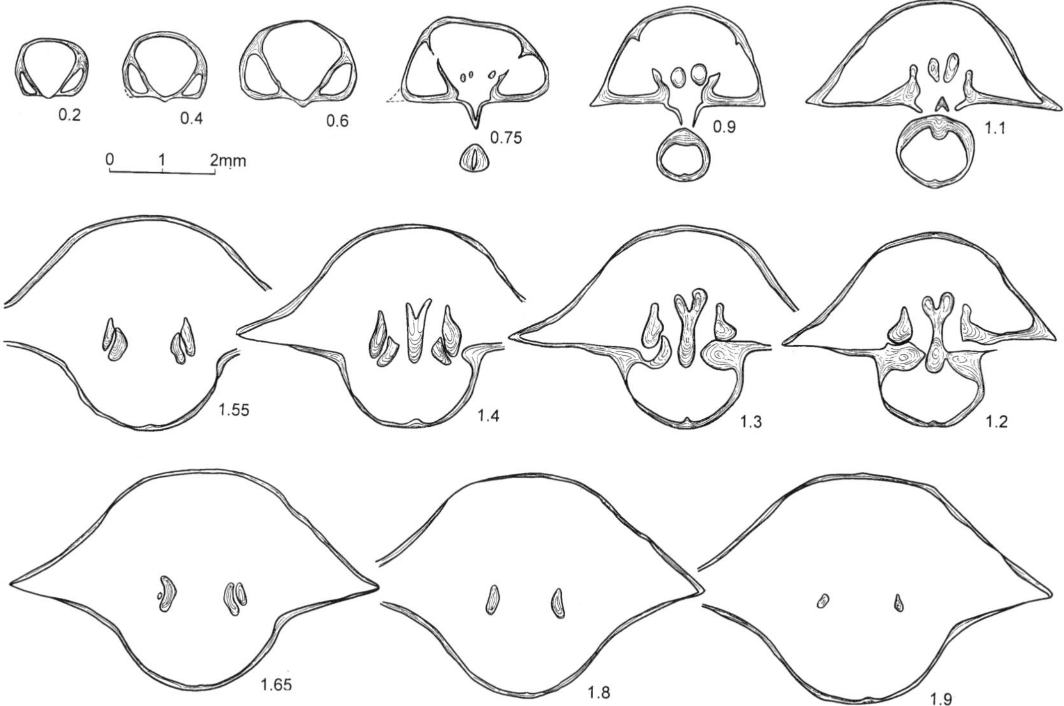

TEXT-FIG. 6. Transverse serial sections of *Triplesia zhejiangensis* Liang, *in* Liu *et al.*, 1983; the positions of the sections are shown on Text-figure 5.

Superfamily TRIPLESIOIDEA Schuchert, 1913
Family TRIPLESIIDAE, Schuchert, 1913

Genus TRIPLESIA Hall, 1859

Triplesia zhejiangensis Liang, *in* Liu *et al.*, 1983

Plate 3, figures 6–11; Text-figures 5–6

1983 *Triplesia zhejiangensis* Liang, *in* Liu *et al.*, p. 271, pl. 97, figs 7–12.

Material. Thirty-nine conjoined valves from the greyish-green calcareous mudstone of the middle to upper part of the Xiazhen Formation at Shiyang of Daqiao, Jiangshan; and 31 moulds, amongst which 12 are dorsal internal and external moulds and 19 are ventral internal and external moulds; mainly from the yellowish-green mudstone of the upper part of the Changwu Formation at Dianbian of Daqiao (300 m from Shiyang of Daqiao), Jiangshan (Locality 6, Collections Jds and Jdd).

Description: exterior. Shells generally 9·0–11·0 mm long, 10·0–12·0 mm wide; length/width ratio *c.* 0·8–0·9. Outline rounded and triangular (Text-fig. 5), and lateral profile dorsi-biconvex. Small epithyridid pedicle foramen. Dorsal fold and ventral sulcus well-developed; the fold originating at about one-third shell length; sulcus generally starting from two-fifths to one-half of the shell length. The sulcus is 40–60 per cent. of shell width at the anterior margin. Anterior commissure strongly uniplicate. Hinge line straight and *c.* 60 per cent. of the maximum width. Ventral interarea high (height H in measurements), apsacline, delthyrium completely covered by a pseudodelthyrial plate with a pseudodeltidial fold in the middle originating from the beak. Dorsal interarea poorly developed.

Ventral interior. Teeth strong; the dental plates thin and short and widely divergent. Muscle field without any bounding ridges; mantle canals bifurcating from the muscle field (Pl. 3, fig. 6).

Dorsal interior. Large cardinal process projecting ventrally and posteriorly, occupying the whole ventral delthyrial cavity with its base near the dorsal beak, shaft long and strong, and myophore bifurcated. Sockets deep; high inner socket ridges connecting with the base and the lower part of shaft of the cardinal process. Myophragm low, originating from the umbo and extending for about 10 per cent. of the shell length (Text-fig. 6).

Measurements.

NIGP Cat. No.	L	W	L/W	T	W_1	W_2	W_3	H
128044 (ventral valve)	6·9	6·3?	1·10	—	3·9	—	—	—
128045 (conjoined valves)	10·7	12·2	0·88	6·6	7·6	7·1	4·9	0·9
128046 (conjoined valves)	10·3	11·4	0·90	6·7	6·4	6·7	3·4	1·3

Remarks. Wright (1993) considered that ornament, shell outline and the pseudodeltidial fold should be used as generic, rather than familial, characters, and we agree. Thus the scheme suggested by Havlíček and Štorch (1990), in which triplesiids were subdivided into three families according to their shell outline and ornament, is not followed here, and all triplesiids are considered as the single family Triplesiidae. The material studied here was collected from the same locality and horizon as the type specimens of *T. zhejiangensis*. But the specimens described by Liang (*in* Liu *et al.* 1983) are larger and have much stronger folds and sulci and more prominent tongues. Thus we have extended Liang's definition of the species to include the wider variation now known from the type locality. Fu (1982, p. 112, pl. 33, fig. 12) identified triplesiids from the mid Ordovician Jinhe Formation at Liquan of Shaanxi as *T.* cf. *extans* (Emmons) and his form is generally similar to our material, but differs in the strongly curved ventral beak covering most of the interarea, and in the distinctive large fold and sulcus which originate from the umbo. Xu (1996, pl 1, figs 26–27) also illustrated a specimen as *Triplesia* sp., from the upper member of the Shiyanhe Formation (middle Ashgill) of Xichuan, Henan Province; this differs from *T. zhejiangensis* in having a much shorter hinge line and wider fold and sulcus. *T. dolborica* Nikiforova (*in* Nikiforova and Andreeva 1961), from the Ashgill of Siberia, differs from *T. zhejiangensis* in having a small median groove on the dorsal fold and a small fold in the ventral sulcus which makes the anterior commissure slightly sulciplicate.

Genus OXOPLECIA Wilson, 1913

Oxoplecia sp.

Plate 3, figure 12

Material. Three dorsal internal moulds, from the yellowish-green mudstone of the top of the Xiazhen Formation at Zhuzhai of Qunli, Yushan (Locality 1, Collection Yz 18).

Remarks. According to Wright (1993, p. 484), the main characters of *Oxoplecia* are a round to elliptical outline, uniplicate anterior commissure, well-developed radial and concentric ornamentation, and delthyrial plates without any median fold. Our material has most of these characters (except that we do not possess any ventral interiors, and thus cannot determine the presence of delthyrial plates): thus we identify it as *Oxoplecia* sp. Our material has characteristic small flaring socket plates, but there are not enough specimens to identify it to species level with confidence.

Superfamily PLECTAMBONITOIDEA Jones, 1928
Family BIMURIIDAE Cooper, 1956

Genus BIMURIA Ulrich and Cooper, 1942

Bimuria? sp.

Plate 3, figure 13

Material. One dorsal internal mould from the yellowish-green mudstone of the upper part of the Changwu Formation at Dianbian of Daqiao, Jiangshan (Locality 6, Collection Jdd).

Remarks. The specimen has a simple cardinal process, a long and highly elevated bema and well-developed side septa, and thus we identify it as perhaps belonging to *Bimuria.* The type species, *B. superba* from the Llandeilo of Tennessee, USA, differs from our specimen in having a larger shell of almost equal length and width, a bema without any postero-lateral ridges, a thin median septum, and no subperipheral rim. If our specimen really is a *Bimuria*, then it would be the first undoubted record from China and one of the first from the Ashgill anywhere, although it is recorded from Scotland (Harper 1989). The *Bimuria* sp. recorded by Liu *et al.* (1983, p. 277) from the early Ashgill Tangtou Formation of Anhui Province, East China, was reassigned to *Christiania* by Cocks and Rong (1989, p. 101).

EXPLANATION OF PLATE 4

Figs 1–5. *Synambonites biconvexus* Rong and Zhan, 1995; Changwu Formation, Dianbian, Locality 6, Horizon Jdd. 1, NIGP 125417; latex cast of ventral external mould; ×5. 2–3, NIGP 124515, holotype; dorsal internal mould; ×3; and latex cast; ×5. 4–5, NIGP 124514; latex cast (×5) and internal mould (×4) of ventral interior.

Fig. 6. Indeterminate xenambonitid gen. et sp. nov.; NIGP 128049; Changwu Formation, near Locality 8, Horizon Jc; dorsal internal mould and part of ventral external mould; ×8.

Fig. 7. *Anoptambonites* sp.; NIGP 128050; Changwu Formation, Locality 8, Horizon Jw; dorsal internal mould; ×5.

Figs 8–13. *Rongambonites bella* gen. et sp. nov.; Changwu Formation, Dianbian, Locality 6, Horizon Jdd. 8, NIGP 128051; ventral internal mould; ×3. 9, NIGP 128052; ventral internal mould; ×3. 10, 12, NIGP 128053; dorsal internal mould and latex cast; ×3. 11, 13, NIGP 128054, holotype; dorsal internal mould (×4) and enlarged view of the cardinalia (11, ×10).

Figs 14–15. *Kassinella* (*Kassinella*) *shiyangensis* sp. nov.; NIGP 128055; Changwu Formation, Shiyang, Locality 6, Horizon Jdy-1; dorsal and ventral views of conjoined valves; ×10.

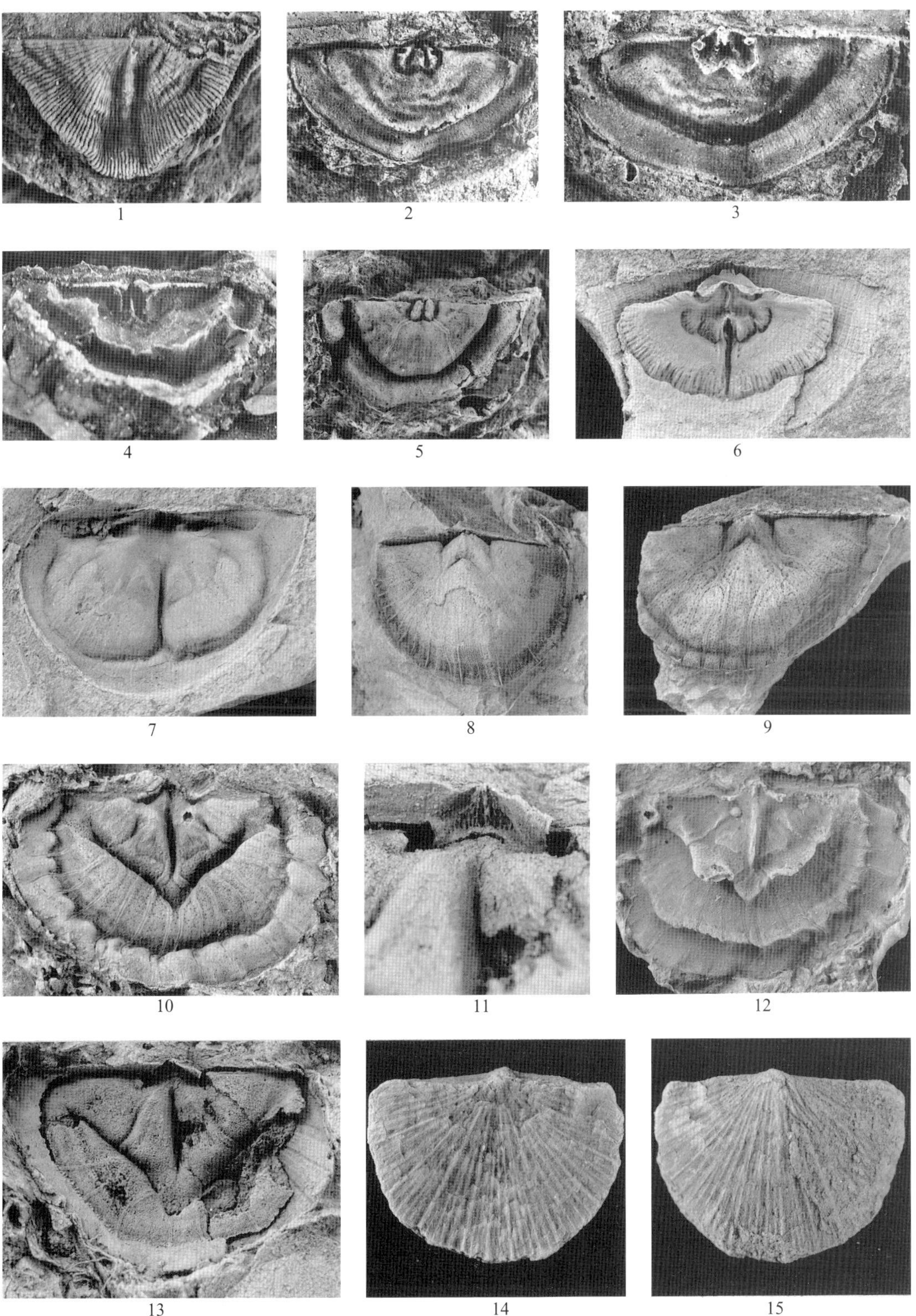

ZHAN and COCKS, late Ordovician brachiopods

Family LEPTELLINIDAE Ulrich and Cooper, 1936

Remarks. The endemic *Reversella trigonoformis* Liang (*in* Liu *et al.* 1983) was allocated to the Leptellinidae by Cocks and Rong (1989), and has its type horizon and locality in the Changwu Formation of Chunan County, western Zhejiang Province, which is about 60 km north-east of our area. However, *R. trigonoformis* has not been found in our substantial collections from the Changwu Formation.

In addition, Liang (*in* Liu *et al.* 1983) described a further new leptellinid, *Qianjiangella qianjiangensis*, from the same formation and locality as *Reversella*; however, this *Qianjiangella* was placed in synonymy with *Leptellina* (*Leptellina*) itself by Cocks and Rong (1989). We have found only a single ventral interior of *L.* (*L.*) *qianjiangensis* within the present study area, from Locality 6, Collection Jdy-1; this is not illustrated here since it adds nothing to Liang's illustrations.

Family XENAMBONITIDAE Cooper, 1956
Subfamily XENAMBONITINAE Cooper, 1956

Genus METAMBONITES Rong and Zhan, *in* Zhan and Rong, 1995

Metambonites meritus Rong and Zhan, *in* Zhan and Rong, 1995

Plate 3, figures 14–17

1995 *Metambonites meritus* Rong and Zhan, *in* Zhan and Rong, p. 564, pl. 1, figs 9, 13; pl. 3, figs 1–12; pl. 4, figs 15–22; text-figs 8–10.

Remarks. This hitherto monospecific genus was described by Rong and Zhan (*in* Zhan and Rong 1995, where detailed information on specimens and localities can be found) from the yellowish-green mudstone of the Changwu Formation, from Dianbian of Daqiao, Jiangshan (Locality 6, Collection Jdd), within our study area. The two genera differ in shape, with *Synambonites* demonstrating alternating convex and concave lateral profiles, whereas *Metambonites* has a simple biconvex profile. In addition, *Metambonites* has a ventral muscle field which is much larger, stronger and more elevated than *Synambonites*, and *Synambonites* has a ventral subperipheral rim which is lacking in *Metambonites*. *Metambonites* was thought by Zhan and Rong (1995) to be endemic to South China. However, some material collected by L. E. Popov from the late Caradoc or early Ashgill Degeres Beds of the Dulankara Formation in the Chu-Ili Mountains of Kazakhstan (BC 12909-10), and which is undergoing systematic revision by I. F. Nikitin and L. E. Popov, can also certainly be assigned to *Metambonites*.

Genus SYNAMBONITES Rong and Zhan, *in* Zhan and Rong, 1995

Synambonites biconvexus Rong and Zhan, *in* Zhan and Rong, 1995

Plate 4, figures 1–5

1995 *Synambonites biconvexus* Rong and Zhan, *in* Zhan and Rong, p. 560, pl. 1, figs 8, 12; pl. 2, figs 1–12; text-figs 6–7.

Remarks. This monospecific genus was described by Rong and Zhan (*in* Zhan and Rong 1995) from the yellowish-green mudstone of the Changwu Formation at Dianbian of Daqiao, Jiangshan (Locality 6, Collection Jdd), which is within our present study area. However, one of the authors (ZRB), together with Prof. Rong Jia-yu and others, found more than 30 specimens from the middle part of the Beiguoshan Formation (middle Ashgill) in Longxian of Shaanxi Province, which lies on the south-west margin of the North China Plate, which are probably *Synambonites* and may be another species. *Xenambonites* can be distinguished from *Synambonites* in having a plano- to concavo-convex lateral profile, a ventral rounded median fold and a corresponding dorsal sulcus, and vestigial dental plates; whereas *Synambonites* has a ventri-biconvex profile, a smaller fold and sulcus, and strong dental plates.

indet. xenambonitid gen. et sp. nov.

Plate 4, figure 6

Remarks. One specimen, including moulds of the interior of both valves and the ventral exterior, from the yellowish-green mudstone of the Changwu Formation at Outangdi of Jiangshan (500 m west of Locality 8), is similar to specimens collected by L. E. Popov from the upper part of the Anderken Formation (middle Caradoc) at Anderkenyn-Akchoku in the Chu-Ili Mountains, Kazakhstan (e.g. BC 12908–129014), which have the same exteriors and interiors except for the denser concentric growth lines in the Kazakhstan specimens. Dr Popov is describing a new genus based on his material, which will be assigned to the Xenambonitidae on the presence of a bema and the undercut cardinal process.

Family HESPEROMENIDAE Cooper, 1956

Genus ANOPTAMBONITES Williams 1962

Anoptambonites sp.

Plate 4, figure 7

Material. One dorsal internal mould from the yellowish-green mudstone of the Changwu Formation at Wujialong of Hejiashan, Jiangshan (Locality 8, Collection Jw).

Remarks. The single specimen is very similar to the type species, *A. grayae* (Davidson) from the upper Caradoc of Scotland (Williams 1962), except that the platform is not so elevated. The median septum is not present anteriorly of the platform and thus this dorsal valve is referred to *Anoptambonites* rather than *Aulie* Nikitin and Popov (*in* Klenina *et al.* 1984). Both genera are known from the upper Ordovician of Russia and Kazakhstan (Klenina *et al.* 1984). Zeng (*in* Wang *et al.* 1987, p. 229) erected the species *A. miner* from the early Caradoc Miaopo Formation of Hubei Province, South China, but we reassign it here to *Kassinella* (*Kassinella*).

Genus RONGAMBONITES gen. nov.

Derivation of name. In honour of Professor Rong Jia-yu.

Type species. *Rongambonites bella* gen. et sp. nov.

Diagnosis. Like *Anoptambonites*, but having a distinctive triangular dorsal platform with straight antero-lateral ridges.

Remarks. Interpretation of the dorsal valve structures is difficult. The strong triangular structure, which is so distinctive as to warrant the erection of a new genus, we consider to represent the platform; however, the two lateral ridges stop short of the hinge line and might in reality represent a bema rather than a platform. The problems are confounded by the faint pair of diverging septa within this structure which could be either side septa or trans-muscle ridges; unfortunately the muscle scars are not sufficiently deeply impressed to be sure. Obviously, if the triangular structure really represents a bema, it would be assigned to the family Xenambonitidae as characterized by Cocks and Rong (1989). However, one of the reasons for interpreting the structure as a platform rather than a bema is the overall similarity of our new genus to *Anoptambonites*; for example the striae seen on the cardinal process of both genera are rare within the Plectambonitoidea. The new genus and *Anoptambonites* differ from *Hesperomena* itself in their strong platforms, dorsal median septa and more prominent teeth, and from other genera in the family in the absence of tubercles in the ventral interior.

Rongambonites bella gen. et sp. nov.

Plate 4, figures 8–13

Derivation of name. From *bella* (Latin): beautiful.

Material. Six ventral and two dorsal internal moulds from the yellowish-green mudstone of the upper part of the Changwu Formation at Dianbian of Daqiao, Jiangshan (Locality 6, Collection Jdd).

Description: exterior. Length 7·3–10·3 mm; width 10·8–17·4 mm; length/width ratio 0·53–0·78. Semi-elliptical to semicircular outline; concavo-convex lateral profile, with ventral valve strongly convex and thickest in the antero-medial part; dorsal concavity variable. Cardinal extremities from acute to round; the maximum width along hinge line. Ventral interarea apsacline and dorsal interarea anacline. Ornament of *c.* 12 costellae, with six or seven intercalated parvicostellae between each pair of costellae.

Ventral interior. Teeth strong; dental plates variably developed, extending parallel to form the lateral bounding ridges of the relatively small muscle field (*c.* 18–20 per cent. shell length and 20–25 per cent. shell width), and with an inverted V-shaped anterior bounding ridge. Both vascular media and vascular myaria originating from the antero-lateral ends of the muscle field, vascular myaria branching twice postero-laterally and antero-laterally.

Dorsal interior. Undercut cardinal process strongly striated posteriorly, connecting laterally with the flaring thick high and triangular socket ridges. Triangular platform with strong and straight antero-lateral bounding ridges at 75–90° to each other, joined anteriorly by a curved and raised section. High median septum not reaching anteriorly to the platform. Muscle field variably impressed, with weak side septa. Low and poorly developed subperipheral rim extending posteriorly to meet the hinge line. Several pairs of evenly distributed pallial markings originating in front of the bema and extending to the shell margin.

Measurements.

NIGP Cat. No.	L	W	L/W	W_4
128051 (ventral valve)	8·4	10·8	0·78	2·5
128052 (ventral valve)	8·5	15·8	0·53	2·9
128053 (dorsal valve)	11·3	17·8	0·63	—
128054 (dorsal valve, holotype)	8·3	13·1	0·63	—

Remarks. The ventral valve is similar to that of *Anoptambonites subcarinatus*, which is from a late Ordovician carbonate mound in central Kazakhstan (Nikitin and Popov 1996, pp. 10–12, fig. 5A–J). However, the distinctive dorsal interior makes *Rongambonites bella* the only species referable to our new genus.

Genus KASSINELLA (KASSINELLA) Borissiak, 1956

Type species. Kassinella globosa Borissiak, 1956, p. 51, pl. 12, figs 1–7; from the upper Ordovician of central to south Khazakhstan.

Emended diagnosis. Shell outline semicircular, plano- or concavo-convex profile. Teeth and dental plates small, ventral median septum high and short; small bilobed ventral muscle field with weak bounding ridges. Cardinal process strong, undercut, connecting with well-developed socket plates laterally (Text-fig. 7); thick and high median septum concave posteriorly in lateral profile;

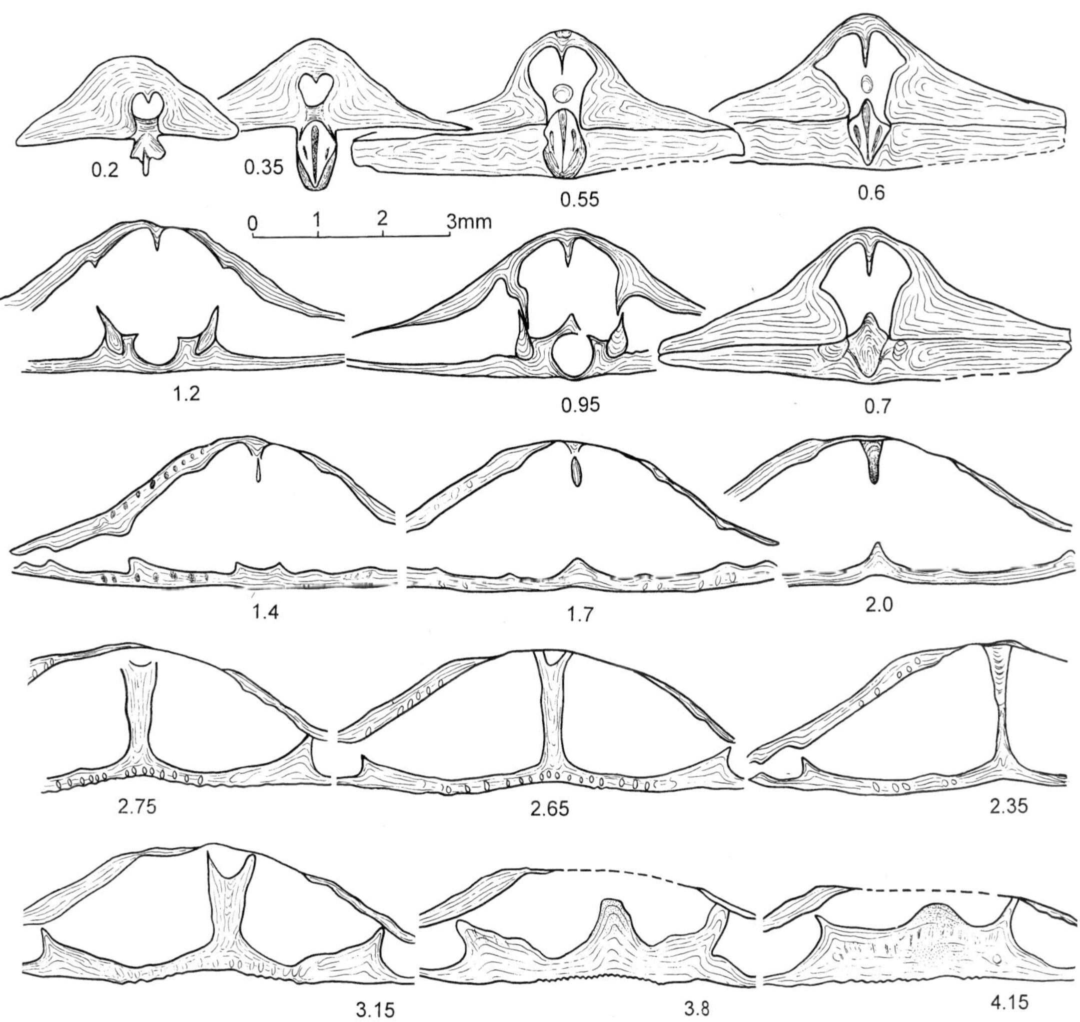

TEXT-FIG. 7. Transverse serial sections of *Kassinella* (*Kassinella*) *shiyangensis* sp. nov. (48 sections made and 16 selected herein) from the Changwu Formation, Locality 6, Collection Jdy-1.

bifurcating at its top end and reaching the ventral floor to enclose the anterior part of the ventral median septum in older specimens, and terminating at the visceral platform. Dorsal muscle field in front of the alveolus, slightly higher than shell floor. Dorsal subperipheral rim present. A pair of tubercles often appears postero-laterally in both valves.

Remarks. Because *Kassinella* is such an abundant constituent of our fauna, we have a much clearer knowledge of its characters. We now know that the chilidium and pseudodeltidium are well developed and slightly convex; the dorsal adductor scars may be slightly elevated, and the dorsal

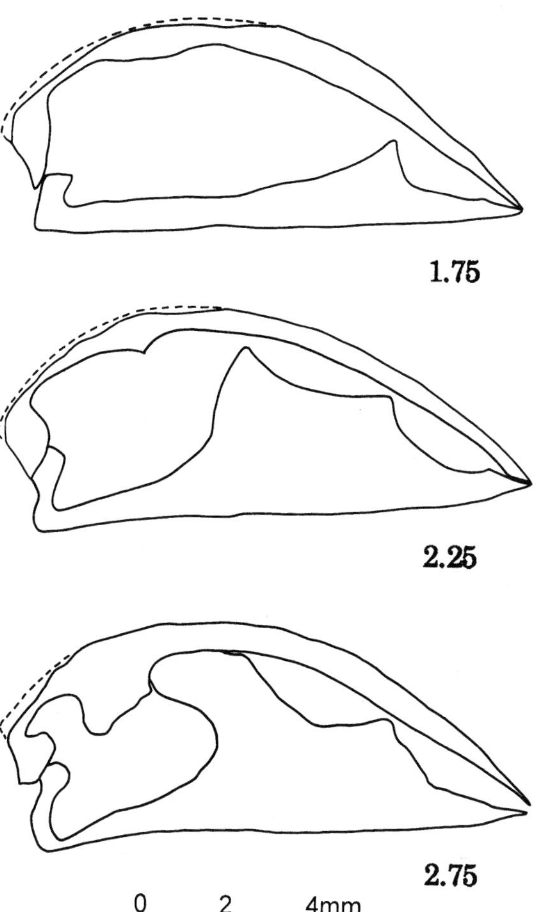

TEXT-FIG. 8. Longitudinal serial sections of *Kassinella* (*Kassinella*) *shiyangensis* sp. nov. (17 sections made and three selected herein) from the Changwu Formation, Locality 6, Collection Jdy-1.

median septum can reach to the ventral floor in adult specimens and even enclose the ventral median septum. Thus we present a fresh generic diagnosis here. *Trimurellina*, originally described from the Ashgill of Ireland (Mitchell 1977) is regarded as another subgenus of *Kassinella* (Cocks and Rong 1989). *K.* (*Kassinella*) is known from the upper Caradoc and Ashgill of Kazakhstan, South China, New South Wales, Scotland, Bohemia and Sweden.

Kassinella (*Kassinella*) *shiyangensis* sp. nov.

Plate 4, figures 14–15; Plate 5, figures 1–7; Text-figures 7–9

1988 *Kassinella* cf. *anisa* Cocks and Rong, p. 60, pl. 9, figs 11–13, 15.
1989 *Kassinella* (*Kassinella*) sp. Cocks and Rong, p. 131, figs 112–114.

Material. One hundred and thirty-eight moulds and 506 individuals with conjoined valves; mainly from the yellowish-brown mudstone (moulds) or greyish-green calcareous mudstone (conjoined valves) of the upper part of the Changwu Formation at Dianbian-Shiyang of Daqiao, Jiangshan (Locality 6, Collection Jdy-1).

Description: Exterior. Shells generally 5·5–7 mm long, 7–8·5 mm wide and 2·3–2·8 mm thick, with a length/width ratio of 0·66–0·91. The smallest known individual is 2·5 mm long, 2·8 mm wide and 1·0 mm thick, whilst the largest is 6·05 mm, 14 mm and 2·65 mm in length, width and thickness respectively. Outline semi-elliptical to semicircular; lateral profile concavo-convex with some dorsal valves nearly flat. Ventral interarea low, apsacline, delthyrium with complete pseudodeltidium; dorsal interarea smaller, anacline, notothyrium with well-developed chilidium. Radial ornament of unequal parvicostellae; ventral valve with 11–14 costellae and parvicostellae between each pair of larger costellae; dorsal valve with 8–11 costellae and 9–12 parvicostellae per 2 mm near the shell anterior margin.

Ventral interior. Teeth strong, dental plates small. Well-impressed bilobed muscle scars, with a thin and high median septum in the postero-median part and a pair of anterior bounding ridges which combine medially with the inserting dorsal median septum to form a strong fulcrum (Text-fig. 8); the central adductor scars small, cone-like; diductor scars as long as 60 per cent. of the shell length, and as wide as 20 per cent. of the shell width. Two faint subperipheral rims, corresponding to the dorsal visceral platform and subperipheral rim, variably developed, the inner one is a little higher and has a series of tubercles along it; the outer one is composed of a series of small tubercles, and is interrupted by pallial markings. A pair of prominent tubercles on the postero-lateral part of the shell. Pseudopunctae coarse and relatively sparse.

Dorsal interior. High and stout undercut cardinal process, connecting with stubby socket plates laterally (Text-fig. 9). Muscle scars small. Median septum high and strong, concave posteriorly in lateral profile, the top

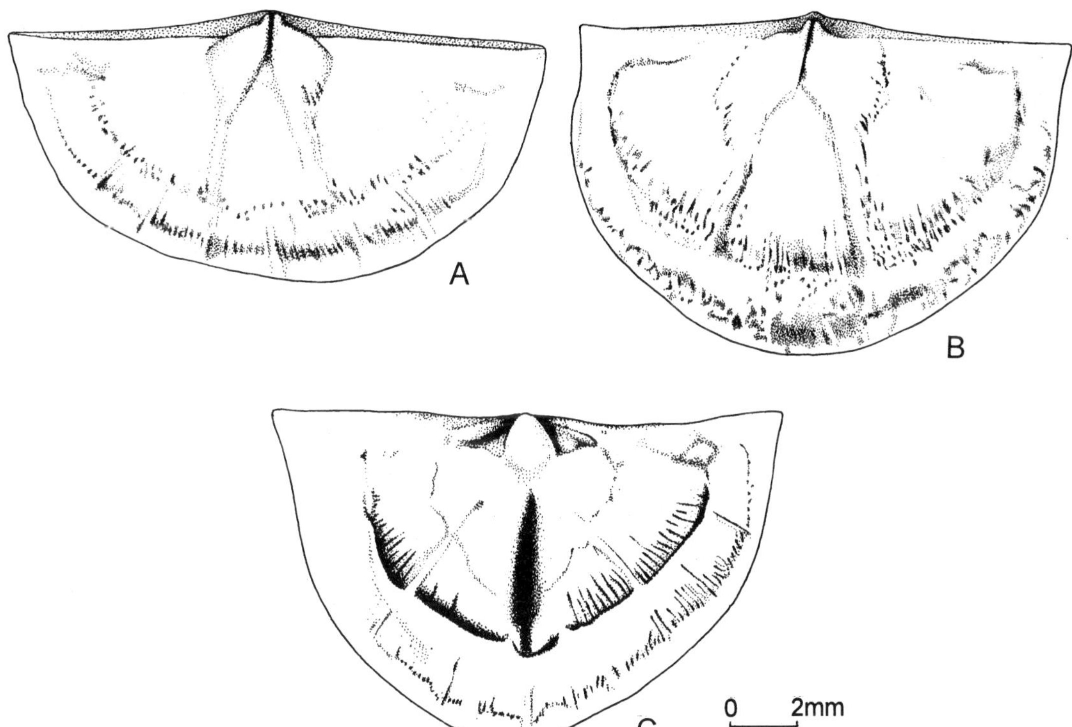

TEXT-FIG. 9. Internal moulds of *Kassinella* (*Kassinella*) *shiyangensis* sp. nov. from the Changwu Formation, Locality 6, Collection Jdy. A, NIGP 128014; B, NIGP 128056 (see Pl. 5, fig. 1); C, NIGP 128059 (see Pl. 5, fig. 4).

bifurcating, and enclosing the central part of the ventral median septum posteriorly; usually terminating at the anterior margin of the visceral platform, but never reaching the subperipheral rim. Visceral platform originating in front of the alveolus and socket plates, extending about 75 per cent. of shell length, with its highest position coinciding with the transverse septum where it reaches or nearly reaches the ventral shell floor. Subperipheral rim low, composed of a series of small tubercles; and usually absent in younger adult individuals. Pallial markings often seen along the margin of visceral platform and subperipheral rim. One or two pairs of tubercles often opposite the corresponding ventral ones and connecting with the posterior end of the visceral platform. Sparse and irregular pseudopunctae.

Measurements.

NIGP Cat. No.	L	W	L/W	W_4
128055 (conjoined valves)	3·1	4·1	0·76	—
128056 (ventral valve)	10·5	13·9	0·76	3·6
128057 (ventral valve)	6·1	9·6	0·64	2·3
128058 (dorsal valve)	?	3·1	?	?
128059 (dorsal valve, holotype)	6·4	7·7	0·83	3·1
128060 (conjoined valves)	6·6	10·2	0·65	—

Remarks. Kassinella (*K.*) *globosa* from the middle Ashgill of Kazakhstan (Borissiak 1956, p. 51, pl. 12, figs 1–7), the type species, is close to the present species. The main differences between them are that *K.* (*K.*) *globosa* has a larger delthyrium and notothyrium, a narrower dorsal interarea, obtuse cardinal extremities, concave dorsal adductor scars, and a poorly developed subperipheral rim. *K.* (*K.*) *anisa* Percival (1979*a*, figs 1B: 2; 5B: 1–3; 7A: 1–8) and *K.* (*K.*) *nana* Klenina (*in* Klenina *et al.* 1984, pl. 7, figs 2, 10; pl. 8, figs 20–27) are also similar to *K.* (*K.*) *shiyangensis*, but *K.* (*K.*) *anisa* has denser primary costellae, 16–20 in the dorsal valve, a slightly catacline dorsal interarea, no dental plates, and a very obvious and continuous subperipheral rim. *K.* (*K.*) *nana* has a smaller shell, often less than 4 mm long, and a length/width ratio of 0·60, a distinctive dorsal sulcus originating from the umbo and extending to the anterior margin of the shell and lower or even absent dental plates. *Anoptambonites miner* from the early Caradoc Miaopo Formation of Hubei Province (Zeng, *in* Wang *et al.* 1987), is reassigned here to *K.* (*Kassinella*), but differs from *K.* (*K.*) *shiyangensis* in its many fewer and undifferentiated costellae and in lacking a subperipheral rim in the dorsal interior. Zeng's figures do not clearly show any postero-lateral tubercles in the ventral interior, but other specimens we have collected from the Miaopo Formation possess tubercles and can be assumed to be variants of Zeng's species.

EXPLANATION OF PLATE 5

Figs 1–7. *Kassinella* (*Kassinella*) *shiyangensis* sp. nov.; Changwu Formation. 1, NIGP 128056; Dianbian, Locality 6, Horizon Jdd; ventral internal mould; ×2·5. 2, 5, NIGP 128057; Locality 8, Horizon Jw 26; internal and external moulds of ventral valve; ×3. 3, NIGP 128058; Locality 9, Horizon Jl 4; dorsal internal mould; ×10. 4, NIGP 128059, holotype; Dianbian, Locality 6, Horizon Jdd; dorsal internal mould; ×4. 6–7, Shiyang, Locality 6, Horizon Jdy-1. 6, NIGP 128055; posterior view of conjoined valves; ×10. 7, NIGP 128060; posterior view of conjoined valves; ×4.

Figs 8–13. *Sowerbyella* (*Sowerbyella*) *sinensis* Wang, 1964; Xiazhen Formation. 8, 11, NIGP 128061; Locality 1, Horizon Yz 18; conjoined moulds; ×2; and enlarged view of the cardinalia; ×10. 9, NIGP 128062; Locality 1, Horizon Yz 7; ventral internal mould; ×2. 10, NIGP 128063; Locality 1, Horizon Yz 7; dorsal internal mould and (lower right) NIGP 128064; mould of ventral exterior; ×2. 12–13, NIGP 128065; Locality 3, Horizon Ys; posterior and lateral views of conjoined valves; ×2.

Figs 14–17. *Sowerbyella* (*Rugosowerbyella*) *dianbianensis* sp. nov.; Changwu Formation, Dianbian, Locality 6, Horizon Jdd. 14, NIGP 128066; ventral external mould; ×4. 15–17, NIGP 128067, holotype. 15–16, external and internal moulds of conjoined valves. 17, oblique view showing cardinalia; ×6.

Figs 18–21. *Holtedahlina sinica* sp. nov.; Changwu Formation. 18–19, NIGP 128070, holotype; Locality 9, Horizon Jl 5; dorsal internal and external moulds; ×2 and ×1·5. 20, NIGP 128071; Locality 8, Horizon Jw; ventral internal mould; ×4. 21, NIGP 128072; Locality 9, Horizon Jl 5; dorsal internal mould; ×4.

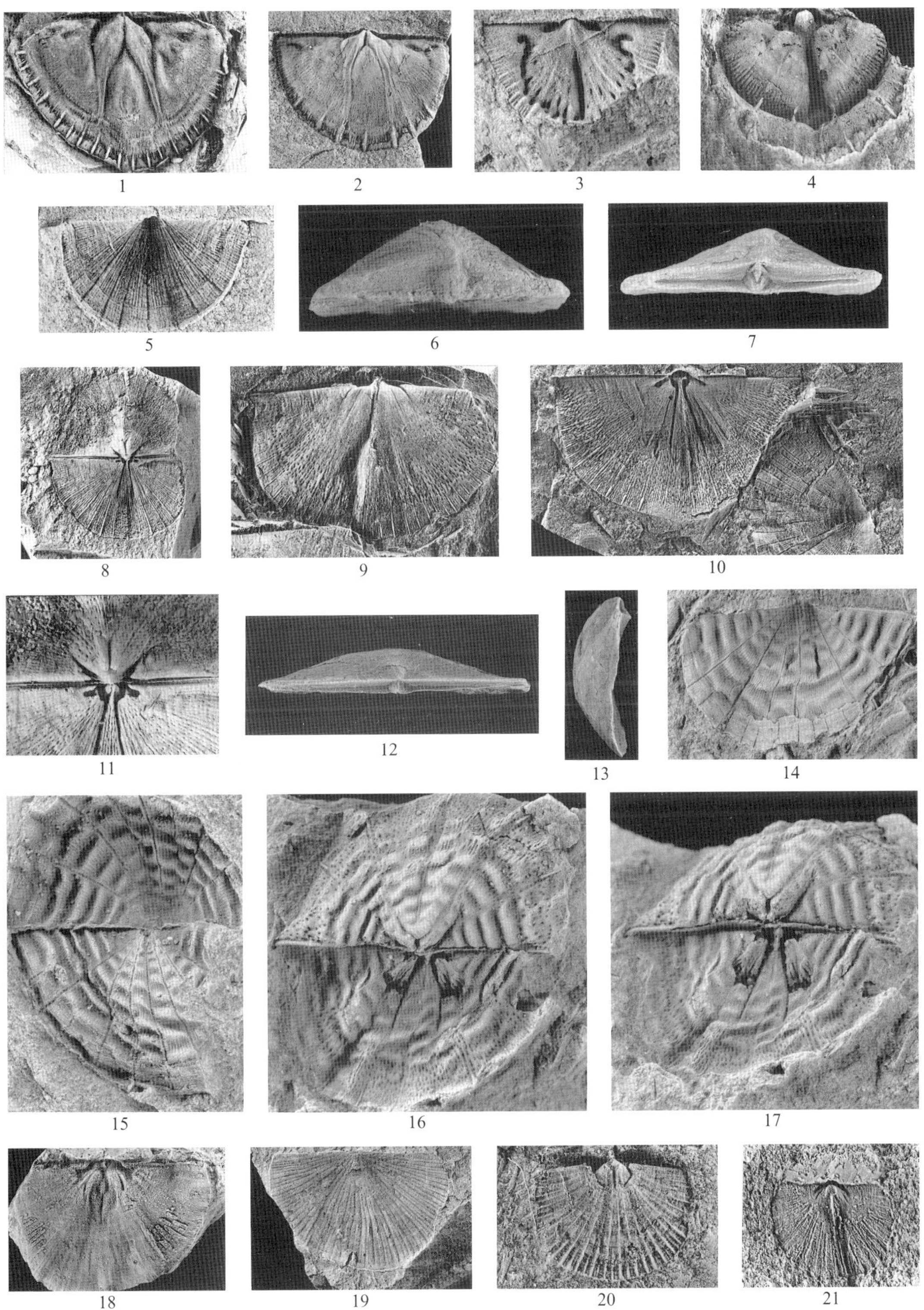

ZHAN and COCKS, late Ordovician brachiopods

Family SOWERBYELLIDAE Öpik, 1930
Subfamily SOWERBYELLINAE Öpik, 1930

Genus SOWERBYELLA (SOWERBYELLA) Jones, 1928

Sowerbyella (*Sowerbyella*) *sinensis* Wang, *in* Wang and Jin, 1964

Plate 5, figures 8–13

1964 *Sowerbyella sinensis* Wang, *in* Wang and Jin, p. 46, pl. 13, figs 9–11.
1983 *Sowerbyella lanxiensis* Liang, *in* Liu *et al*., p. 276, pl. 99, figs 1–3.

Material. One hundred and three specimens with conjoined valves mainly from the greyish-green calcareous mudstone of the lower part of Xiazhen Formation at Shiyanshan of Xiazhen, Yushan (Locality 3, Collection Ys); 1381 moulds amongst which 89 with both valves preserved together, mainly from the yellowish-green mudstone of the Xiazhen Formation at Zhuzhai of Qunli, Yushan, and from the Changwu Formation at Mulinlong of Tanshi, Jiangshan (Localities 1 and 7, Collections Yz 7, Yz 18, Jm 18, Jm 23, Jm 24, and Jm 35).

Description: exterior. Shells generally 6·1–11·6 mm long, 11·3–18·25 mm wide and with a length/width ratio of 0·54–0·73; semi-elliptical in outline and concavo-convex in lateral profile with the thickest position at about the mid-length. Maximum width along hingeline or slightly anterior of it. Cardinal extremities round, rarely extending to form ears. Low, triangular ventral interarea (H in measurements) apsacline, slightly concave, with some transverse fine grooves; delthyrium covered by pseudodeltidium in its posterior half. Low dorsal interarea hypercline; notothyrium covered by small chilidium. Anterior commissure rectimarginate. Ornament of concentric growth lines and differentiated radial costellae. Ventral valve with 23–29 evenly distributed coarser costellae with 3–5 parvicostellae between them; dorsal valve with 11–13 unevenly distributed coarser costellae with 5–14 finer parvicostellae between them; usually 11 costellae per mm anteriorly. Concentric fila dense, a few specimens with weak discontinuous concentric rugae posterolaterally.

Ventral interior. Teeth strong and grooved; prominent dental plates extending antero-laterally to form the low bounding ridges of the muscle field. Bilobed muscle scars well impressed, about 35 per cent. of the shell length and 36 per cent. of shell width; a pair of deep kidney-like small adductor scars in the posterior part of the delthyrial cavity. Low and short median septum continuous anteriorly with the low muscle bounding ridges. A prominent pair of vascular media. Vascular myaria inserted into the muscle field posteriorly, and branching and extending anteriorly to the shell anterior margin. Evenly distributed setal grooves can be seen near the shell anterior.

Dorsal interior. Strong undercut cardinal process highly projecting ventrally and occupying 70 per cent. of the whole delthyrial cavity. Socket ridges thick but short, connected to the cardinal process from both sides, forming an inverted V-shaped structure. On both sides of the alveolus a pair of low ridges originates from the middle of the socket ridges bounding the alveolus; the ridges are continuous with the side septa anteriorly. Median septum absent; median side septa high and thin, starting in front of the alveolus, extending anteriorly with a very small angle ($< 15°$) and ending abruptly at the mid-length, almost equal or slightly shorter than the muscle field. Muscle scars bilobed, sub-triangular, without any lateral bounding ridges, slightly longer than half shell length, but as wide as half the shell. Pallial markings poorly impressed; vascular media within the median side septa, vascular myaria inserted into the muscle field posteriorly and branching anteriorly.

Measurements.

NIGP Cat. No.	L	W	L/W	T	H
128061 (conjoined moulds)	6·6	12·8	0·52	—	—
128062 (ventral valve)	13·2	21·0	0·65	—	—
128063 (dorsal valve)	12·1	21·2	0·57	—	—
128064 (ventral valve)	11·3	19·5	0·58	—	—
128065 (conjoined valves)	12·4	21·8	0·57	3·4	0·8

Remarks. The type horizon of *S*. (*S*.) *sinensis*, originally termed the Sanqushan Formation by Sheng (1951), is now termed the Xiazhen Formation, and most of our material comes from the type locality, Shiyanshan of Xiazhen. *S. lanxiensis* was described from the Changwu Formation by Liang (*in* Liu *et al*. 1983). These two formations are correlatives of each other and the two type localities are about 50 km apart. Liang based his new species on a different valve outline from that of *S*. (*S*.)

sinensis, but our large collections show variability in outline and we synonymize the two species here.

The main differences between *S.* (*S.*) *sinensis* and the type species *S.* (*S.*) *sericea* are the lesser convexity, smaller ventral umbo which seldom projects over the hingeline, the smaller number of parvicostellae between each pair of costellae, and the longer ventral muscle field which often exceeds half of the shell in *S.* (*S.*) *sericea*. *S. nativa* Klenina, from the middle Ordovician of north-eastern Kazakhstan (Klenina *et al.* 1984, p. 78, pl. 6, figs 4–6, 8–9; pl. 7, fig. 11) is also similar to *S. sinensis*, but the former has stronger convexity (and correspondingly greater concavity in the dorsal valve), a flat ventral interarea, very fine and numerous parvicostellae (38–44 costellae and about 172 parvicostellae along the anterior margin), and ventral median septum extending to one-third of the shell length. Other species similar to *S.* (*S.*) *sinensis* include *S. lepta* from New South Wales, Australia (Percival 1979*a*, p. 108, figs 3B, 4–7; 8), *S.* cf. *monilifera* (Mitchell 1977, p. 80, pl. 16, figs 1–8) from Pomeroy, Northern Ireland and *S. thraivensis* (Reed) from Girvan, Scotland (Harper 1989, p. 112). In addition to other differences, including smaller muscle fields, *S. lepta* has ventral valve canals and can thus be assigned to the subgenus *S.* (*Eochonetes*) as reviewed by Cocks and Rong (1989). *S.* (*S.*) cf. *monilifera* of Mitchell can be distinguished from *S.* (*S.*) *sinensis* by the flat ventral interarea, the anacline dorsal interarea, the shallow and wide dorsal sulcus originating near the umbo, the relatively few costellae (only six to nine at the anterior part of the shell), and the long dorsal adductor scar which extends to about two-thirds of the shell length and width. *S.* (*S.*) *thraivensis* differs from *S.* (*S.*) *sinensis* in its flat dorsal valve, catacline dorsal interarea, cordate ventral muscle field about one-half of shell width, degenerate dental plates, very small cardinalia only about 12 per cent. of the shell length and 29 per cent. of the shell width, and anteriorly elevated median side septa.

Specimens from the upper member of the Shiyanhe Formation (middle Ashgill) of Xichuan, Henan Province, were also identified as *S.* (*S.*) *sinensis* by Xu (1996, p. 554, pl. 2, figs 5–13; pl. 4, fig. 8), but they are different from both the type specimens and the present material in having a larger ventral muscle field and a pair of weak, invariably present dorsal transmuscle ridges. Their identification should be reviewed.

Genus SOWERBYELLA (RUGOSOWERBYELLA) Mitchell, 1977

Sowerbyella (*Rugosowerbyella*) *dianbianensis* sp. nov.

Plate 5, figures 14–17

Material. One pair of conjoined valves, one pair of conjoined moulds, three dorsal internal and five external moulds, two ventral and three external moulds, mainly from the yellowish-green mudstone of the upper part of the Changwu Formation at Dianbian of Daqiao, Jiangshan (Locality 6, Collection Jdd).

Description: exterior. Shell 4·0–5·4 mm long, 6·5–9·4 mm wide; length/width ratio 0·57–0·61; semicircular outline; concavo-convex lateral profile with thickest position postero-medially. Cardinal extremities acute, ears well-developed. Anterior commissure rectimarginate or slightly unisulcate with a weak dorsal sulcus originating 3 mm from the umbo. Seven or eight costellae originating from the umbo with about 12 parvicostellae between each pair of them. Five to seven concentric rugae covering all the shell surface, which are usually interrupted by the costellae.

Ventral interior. Small teeth supported by weak divergent dental plates which extend antero-laterally to form the outer bounding ridges of the muscle field. Small adductor scars at the postero-medial part divided by a thin and high myophragm which bifurcates into the inner bounding ridges of the field with an angle of 70°; divergent diductor scars elongate.

Dorsal interior. Cardinalia about 11 per cent. and 18 per cent. of the shell length and width. Strong cardinal process undercut, projecting ventrally, and connecting with socket ridges laterally which are well-developed and 98° in angle and extending antero-laterally to enclose the notothyrial cavity. Bilobed bema variably developed, highly elevated and undercut anteriorly, and with several longitudinal grooves on its antero-medial part. A pair of side septa originating in front of the notothyrial cavity, diverging at 35° and extending about 70 per cent. of the shell length, with the posterior parts forming the inner bounding ridges of the bema.

Measurements.	NIGP Cat. No.	L	W	L/W
	128066 (ventral valve)	5·8	9·7	0·60
	128067 (conjoined moulds, holotype)	4·4	7·8	0·56

Remarks. Mitchell (1977) established *Rugosowerbyella* as a subgenus of *Sowerbyella* from the upper Ordovician of Pomeroy, Northern Ireland, on the basis of regular concentric rugae cut by radial costella. The type species is *S.* (*Rugosowerbyella*) *ambigua* (Reed) (Mitchell 1977, p. 83, pl. 16, figs 23–27) which Cocks and Rong (1989, p. 145) considered a junior subjective synonym of *S.* (*R.*) *subcorrugatella* (Reed) and which can be distinguished from our material in being relatively wider, and in having more costella, and less prominent but more numerous rugae, a less strongly impressed ventral muscle field, and a relatively longer bema. *S.* (*Rugosowerbyella*) has been reported from Scotland (lower Ashgill), Nevada (upper Llanvirn), Kazakhstan (Caradoc), Sweden (lower Ashgill; Cocks and Rong 1989) and Taimyr (Cocks and Modzalevskaya 1997). Our new species is the first record of *S.* (*Rugosowerbyella*) from China.

Superfamily STROPHOMENOIDEA King, 1846
Family STROPHOMENIDAE King, 1846

Genus STROPHOMENA de Blainville, 1824

Strophomena (*Strophomena*) sp.

Plate 6, figures 3–4

Remarks. Six conjoined valves from the greyish-green calcareous mudstone of the lower part of the Xiazhen Formation at Shiyanshan of Xiazhen (Locality 3, Collection Ys) and numerous variably preserved moulds from the yellowish-green mudstone of the upper part of the Xiazhen Formation at Zhuzhai of Qunli, Yushan (Locality 1, Collections Yz 7 and Yz 18), have many similarities with the revised *Strophomena* (*Strophomena*) (Rong and Cocks 1994). Rong and Han (1986) recorded *Strophomena* sp. from the mudstone of the Xiazhen Formation at Zhuzhai of Yushan. The present form is variable in length-width ratio, and differs from the type species, *S.* (*S.*) *planumbona* from the upper Ordovician of Kentucky, USA, in having a relatively smaller ventral muscle field and finer ornament. No previous records exist of *S.* (*Strophomena*) from the Ordovician of South China; however, the genus has been recorded from the Silurian of Guizhou and Hubei provinces (Rong and Yang 1981), but those forms would be more appropriately placed within *Katastrophomena*.

EXPLANATION OF PLATE 6

Figs 1–2. *Holtedahlina sinica* sp. nov.; Changwu Formation. 1, NIGP 128071; Locality 8, Horizon Jw; latex cast of ventral internal mould; ×7. 2, NIGP 128070, holotype; Locality 9, Horizon Jl 5; latex cast of dorsal internal mould; ×3.

Figs 3–4. *Strophomena* (*Strophomena*) sp. 3, NIGP 128068; Changwu Formation, Dianbian, Locality 6, Horizon Jdd; dorsal internal mould; ×2. 4, NIGP 128069; Xiazehn Formation, Locality 1, Horizon Yz 18; ventral internal mould; ×1·5.

Figs 5–13. *Fenomena distincta* gen. et sp. nov.; Changwu Formation, Dianbian, Locality 6, Horizon Jdd. 5–6, NIGP 128073; ventral internal mould and latex cast; ×2. 7–8, NIGP 128074, holotype; dorsal internal mould and latex cast; ×4. 9, NIGP 128075; dorsal internal mould; ×4. 10–12, NIGP 128076; 10–11, dorsal internal mould and latex cast; ×4; 12, details of cardinal area; ×10. 13, NIGP 128077; latex cast of dorsal external moulds with the encrusting cyclostome bryozoan *Corynotrypa* cf. *delicatula* on the shell surface (identified by Dr P. D. Taylor); ×4.

Figs 14–15. *Tashanomena variabilis* Zhan and Rong, 1994; Xiazhen Formation, Locality 2, Horizon Yt 12–4. 14, NIGP 121542; ventral internal mould; ×10. 15, NIGP 121553; dorsal internal mould; ×10.

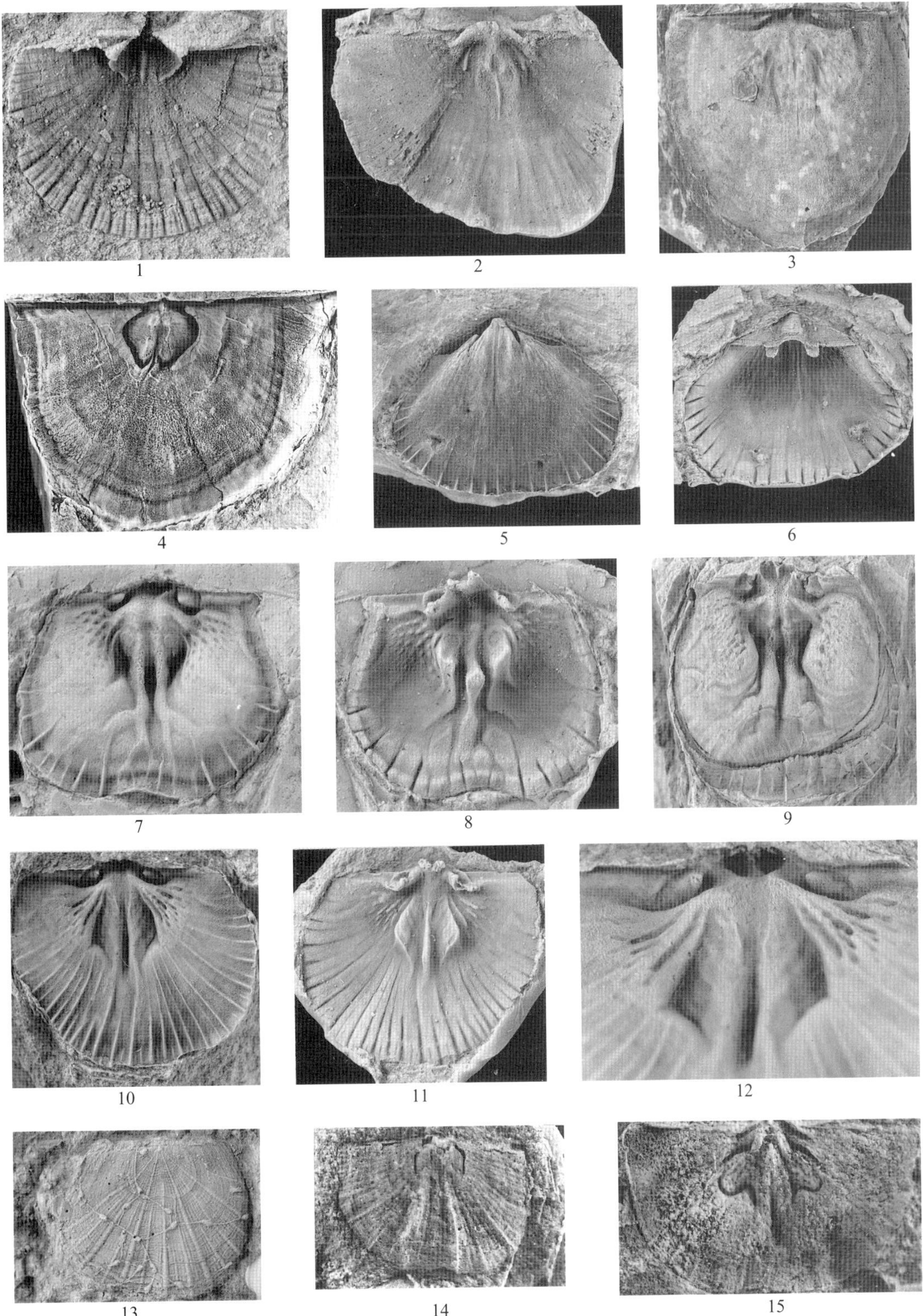

ZHAN and COCKS, late Ordovician brachiopods

Genus HOLTEDAHLINA Foerste, 1924

Type species. Leptaena sulcata de Verneuil, 1848, p. 350, pl. 4, fig 4; from the Waynesville Member of the Richmond Formation (upper Caradoc–lower Ashgill), Indiana and Ohio, USA.

Remarks. Holtedahlina is similar to *Strophomena* in its cardinalia and ventral interiors but its biconvex shell and well-developed dorsal fold and ventral sulcus easily differentiates it. Other strophomenines with biconvex shells are *Infurca, Esilia* and *Trotlandella. Infurca* (Percival 1979*b*, p. 185) is similar to *Holtedahlina* in its cardinalia, but has crenulate teeth supported by divergent dental plates with short bounding ridges to the pentagonal muscle field, which is open anteriorly, and a thin median septum in the dorsal valve. *Esilia* Nikitin and Popov (1985, p. 38) can be distinguished from *Holtedahlina* in its strong dorsal fold and ventral sulcus, elliptical ventral muscle bounding ridges; robust cardinal process, with lobes close together and situated on a strong notothyrial platform; socket ridges curving posteriorly and extending laterally very close to the hinge line. *Trotlandella* Neuman (*in* Neuman and Bruton 1974, p. 95) has delicate cardinal process lobes with the short strong socket ridges curved laterally subparallel to the hinge line.

Holtedahlina sinica sp. nov.

Plate 5, figures 18–21; Plate 6, figures 1–2

Material. Twenty-three dorsal internal and 15 external moulds, eight ventral internal and four external moulds; mainly from the yellowish-green mudstone of the upper part of the Changwu Formation at Pengli of Hejiashan, Jiangshan (Locality 9, Collections Jl 4 and Jl 5).

Description: exterior. Shell 11·0–14·0 mm long, 16·0–20·0 mm wide, length/width ratio 0·6–0·8. The smallest specimen is a dorsal valve 4·0 mm long and 4·96 mm wide; the largest, also a dorsal valve, is 14·0 mm long and 20·7 mm wide. Nearly semicircular outline and biconvex lateral profile. Cardinal extremities round, maximum width at one-third shell length. Low dorsal fold and shallow ventral suclus originating at two-thirds valve length. Apsacline ventral interarea 0·2–0·25 mm high. Dorsal interarea anacline. Ornament of 3–4 parvicostellae between each pair of costellae. Fine concentric growth lines 15–20 per mm.

Ventral interior. Teeth strong; thin and high dental plates extending anteriorly. Bilobed muscle field about one-fifth and one-quarter of shell width and length respectively. Small elliptical adductor scars just in front of pedicle callist, enclosed by diductor scars. Thin and high myophragm slightly longer than the muscle field.

Dorsal interior. Strong bilobed cardinal process with knob-like myophore projecting ventrally, connecting laterally at its base with socket ridges which are thick and short, divergent at 110° to 140° and which extend postero-laterally to join the hinge line; the socket ridges with variably developed crenulations postero-laterally. The socket ridges extend to *c.* 25–30 per cent. of shell width. Adductor scars poorly impressed, with no bounding ridges, but sometimes with a low and wide myophragm which extends anteriorly to form a thin and high median septum.

Measurements. Distance between two lateral ends of socket ridges (W_5) and their divergent angle ($\propto$).

NIGP Cat. No.	L	W	L/W	L_1	W_1	W_5	$\propto$
128070 (dorsal valve, holotype)	11·9	17·5	0·68	4·3	3·4	4·4	112°
128071 (ventral valve)	5·0	6·4	0·79	1·2	1·4	—	—
128072 (dorsal valve)	3·8	4·9	0·77	1·5	1·4	1·6	109°

Remarks. All the specimens are well preserved and apparently not transported. Variations are recognized in: (1) dorsal convexity: among the 23 dorsal internal moulds, five are nearly flat, ten are as convex as the ventral valve, and eight are more convex than the average convexity of the ventral valve, the most convex valve is 4 mm thick; (2) the development of the fold and sulcus: the development of the fold is generally consistent with the dorsal convexity, and also has some relationship to shell size, i.e. during the ontogeny of an individual, the fold changes; in the young stage, the shell is small, with the dorsal valve almost flat, and sometimes even with a shallow groove in the antero-middle part; when the shell is a little larger, the dorsal valve is apparently gently convex; in adult stages the dorsal valve is more convex than the ventral valve, with the median fold comparatively high and wide; (3) dorsal muscle field and myophragm: only one dorsal valve has an obvious thin and high myophragm, four have comparatively smooth muscle fields in which the myophragm is not clear or even absent. The largest specimen has very weak muscle scars, but with a low median ridge; (4) the angle between the socket ridges: this becomes larger during ontogeny but when it reaches the adult stage the angle is still smaller than *Strophomena*, whose socket ridges are nearly parallel to the hingeline.

The type species, *H. sulcata*, can be distinguished from our new species in having a much stronger fold and sulcus, more robust and erect cardinal process, less curved socket ridges and a relatively larger ventral muscle field. *Holtedahlina* is only reliably reported from North America (Foerste 1924; Cooper and Kindle 1936). The *Holtedahlina* sp. from the Dolbor Horizon (Ashgill) of Siberia (Nikiforova and Andreeva 1961) is only doubtfully referred to the genus, since good interiors are not illustrated.

Subfamily FURCITELLINAE Williams, *in* Williams *et al.*, 1965

Genus FENOMENA gen. nov.

Type species. Fenomena distincta gen. et sp. nov.

Diagnosis. Rectangular to sub-semicircular outline; plano-convex to slightly biconvex profile. Large pseudodeltidium; small chilidium. Dental plates short and weak with no ventral muscle bounding ridges. Strong bilobed cardinal process continuous with the socket ridges laterally. Short sockets with strong curved socket ridges. Subperipheral rim in adults. Elevated dissected dorsal muscle field and a dorsal median septum.

Remarks. The cardinalia are of Type A (Rong and Cocks 1994), which, together with the dorsal muscle field, includes this new genus within the Furcitellinae. *Fenomena* is closest to *Dactylogonia* Ulrich and Cooper, 1942 and its junior synonym *Cyphomena* Cooper, 1956, but differs in (1) convexity: *Fenomena* is plano-convex to slightly biconvex whilst *Dactylogonia* is concavo-convex and dorsally geniculate; (2) ventral muscle bounding ridges, which are lacking in *Fenomena* and present in *Dactylogonia*; (3) the side septa, which are always present in *Dactylogonia*, but absent and replaced by the elevated dissected dorsal muscle field in *Fenomena*, and (4) the strong dorsal median septum present in *Fenomena*, but which is absent or only very weakly developed in *Dactylogonia*. The new genus appears to be related to *Oepikina* Salmon, 1942, but this again has a concavo-convex profile with weak geniculation and strongly developed dorsal side septa (Rong and Cocks 1994, pl. 2, fig. 8). *Fenomena* is so far only known from its type species, *F. distincta* from South China.

Fenomena distincta gen. et sp. nov.

Plate 6, figures 5–13

Material. Three ventral internal, 18 dorsal internal and five external moulds from the yellowish-green mudstone of the upper part of the Changwu Formation at Dianbian of Daqiao, Jiangshan (Locality 6, Collection Jdd).

Description: exterior. Shell 6·6–13·9 mm long, 7·9–18·0 mm wide within length/width ratio of 0·76–0·85 (apart from one specimen nearly as wide as long). Outline rectangular or sub-semicircular; lateral profile plano-convex or slightly biconvex. Anterior margin usually round but with occasional weak bilobation anteriorly. Maximum width at shell midlength or slightly anterior of it. Large apsacline ventral interarea with dense fine striae parallel to the delthyrium, which is covered by a complete pseudodeltidium. Small anacline dorsal interarea with a well-developed chilidium. Ornament of 4–6 costellae originating from the umbo whilst 11–15

SOMERVILLE COLLEGE
LIBRARY

costellae are inserted between 25–35 per cent. shell length, 3–5 parvicostellae between each pair of costellae near the anterior margin. Faint but clear concentric fila evenly distributed, 43 per mm.

Ventral interior. Small teeth with several crenulations on their posterior surface, supported by weak dental plates posteriorly which extend forward for only a short distance. Muscle field poorly impressed, long and narrow adductor scars slightly elevated with a weak median septum extending to the shell midlength. Muscle bounding ridges absent. A series of shallow grooves along the shell antero-lateral margin reflects the exterior costellae.

Dorsal interior. Strong bilobed cardinal process lobes fused together at their base and shafts, forming a highly projecting plate-like structure which continues into the socket ridges laterally. Small and deep sockets surrounded by socket ridges, weakly developed but thick posterior ridges with some crenulations on their inner surface. Chief anterior ridges high but thin, curved postero-laterally, and extending parallel to the hinge line and connecting with subperipheral rim at their lateral ends in older specimens. Muscle field becoming progressively elevated anteriorly: this is dissected and performs the same function as thickened side septa. Variably developed median septum, highest near the anterior margin of the muscle field and continuing to up to four-fifths shell length. A pair of deep, parallel grooves separates the median septum and the elevated muscle field. Many small tubercles lateral to the muscle field on both sides. Low subperipheral rim present in adult specimens, cut across by a series of small grooves antero-laterally, reflecting the exterior costellae. Pallial markings lemniscate; a pair of vascular media originating in front of the muscle field, branching three times; vascular myaria starting at the medial parts of muscle bounding ridges, extending antero-laterally and branching twice.

Measurements.

NIGP Cat. No.	L	W	L/W
128073 (ventral valve)	14·0	17·7	0·79
128074 (dorsal valve, holotype)	8·5	10·4	0·82
128075 (dorsal valve)	12·5	13·2	0·95
128076 (dorsal valve)	8·0	10·1	0·79
128077 (dorsal valve)	5·8	7·4	0·78

Remarks. No similar species have been reported from the upper Ordovician in neighbouring or distant areas, and it is therefore believed to be endemic to our area.

Family GLYPTOMENIDAE Cooper, 1956
Subfamily TERATELASMINAE Pope, 1976

Genus TASHANOMENA Zhan and Rong, 1994

Tashanomena variabilis Zhan and Rong, 1994

Plate 6, figures 14–15

1994 *Tashanomena variabilis* Zhan and Rong, p. 416, pl. 1, figs 1–17; text-figs 2–6.

Remarks. Zhan and Rong erected the monospecific genus *Tashanomena* from the present fauna, with its type horizon and locality from the middle part of the Xiazhen Formation at Tashan of Xiazhen, Yushan (Locality 2, Collection Yt 12–4), to reflect those teretelasmines with oblique postero-lateral and transverse S-shaped antero-lateral dorsal muscle bounding ridges and variably developed dorsal median septum. As far as is known, both the genus and its type species are endemic to Southeast China.

indet. teratelasmine gen. et sp. nov.

Plate 7, figures 1–3

Remarks. One ventral internal mould and one dorsal valve with internal and external moulds, from the yellowish-green mudstone of the upper part of the Changwu Formation at Dianbian of Daqiao, Jiangshan (Locality 6, Collection Jdd), represent a new genus and species, but the material is at present insufficient to erect these taxa formally. The dorsal valve has small separated cardinal process lobes with nearly straight socket ridges, and also side septa, which allocates it to the subfamily Teratelsaminae according to Rong and Cocks' (1994) revision. *Teretalasma* differs from the Chinese form in having well-developed dorsal transmuscle ridges and a dorsal median septum. *Tashanomena*, another endemic teratelasmine in the present fauna, can be distinguished from it by having a concavo-convex shell, transverse S-shaped antero-lateral dorsal muscle bounding ridges and a variably developed dorsal median septum.

Family FOLIOMENIDAE Williams, 1965

Genus FOLIOMENA Havlíček, 1952

Remarks. Havlíček (1952) erected *Foliomena*, with *Strophomena folium* as its type species, for flat transverse shells without radial ornament; no dental plates; small bilobed cardinal process; long socket plates extending laterally; and a pair of side septa. We can add to the diagnosis of *Foliomena* in observing that the dorsal side septa become higher and thicker anteriorly with a shallow median groove between them.

Foliomena folium (Barrande, 1879)

Plate 7, figure 10

1879 *Strophomena folium* Barrande, pl. 55, I, figs 1–13.
1952 *Foliomena folium* (Barrande, 1879); Havlíček, p. 17, pl. 3, figs 1, 3–4, 6.
1988 *Foliomena folium* (Barrande 1879); Cocks and Rong, p. 53, pl. 9, figs 1, 7–8, 14, 16.

Material. Twenty-six moulds with conjoined valves, 58 ventral and 61 dorsal moulds; mainly from the greyish-green mudstone of the upper part of the Changwu Formation at Shangwu of Fengzu, Jiangshan (Locality 10, Collection Js 2).

Remarks. Our material is from the same locality and horizon as one of the specimens figured by Cocks and Rong (1988, pl. 9, fig. 16). The type species, *F. folium* (Barrande, 1879) from Bohemia, is now the only reliable species within a narrow definition of *Foliomena.* Cocks and Rong (1988, p. 65) identified a specimen from the late Ordovician Crugan Mudstone of northern Wales as *Foliomena*? sp. because it has a clear central radial costella. In addition, Harper (1989) erected *F. exigua* from the Ashgill of Girvan, Scotland, with a single central costella, and Sheehan and Lespérance (1979) described *F. joliensis* from the Ashgill of Percé, Quebec, Canada, which has three radial costellae. We think that these ribbed forms would be better attributed to a separate subgenus of *Foliomena*, one that is not known from China. *Foliomena folium* is a central constituent of the widespread *Foliomena* fauna, which is relatively cosmopolitan and inhabited deeper areas of the shelf and the slope (Cocks and Rong 1988).

Family CHRISTIANIIDAE Williams, 1953

Genus CHRISTIANIA Hall and Clarke, 1892

Christiania aff. *magna* Sheehan, 1987

Plate 7, figures 4–6, 8–9

aff. 1987 *Christiania magna* Sheehan, p. 37, pl. 11, figs 31–32; pl. 12, figs 1–17.

Material. Five conjoined valves from the greyish-green calcareous mudstone of the upper part of the Changwu Formation at Mulinlong of Tanshi, Jiangshan; nine dorsal internal moulds and one external mould, and six ventral internal and three external moulds, from the yellowish-green mudstone of the upper part of the Changwu Formation at Dianbian of Daqiao, Jiangshan (Locality 6, Collection Jdd).

Remarks. Our material is very close to *C. magna* from the Ashgill of Belgium (Sheehan 1987) but differs from the latter in having a slightly more transverse shell and a deeper dorsal valve. In addition, the raised anterior bounding ridges of the dorsal muscle field between the two pairs of side septa appear to be less parallel to the hinge line than in the Belgian form. However, we do not yet have sufficient data to erect a new species, and thus here identify the Chinese form as *C.* aff. *magna.* The *Bimuria* sp. of Liu *et al.* (1983, p. 277, pl. 92, fig. 17), which was reassigned to *Christiania* by Cocks and Rong (1989), and the *Christiania* cf. *subquadrata* (Hall) of Liu *et al.* (1983, p. 277, pl. 92, fig. 21), are both from the early Ashgill Tangtou Formation at Chuxian of Anhui Province, East China, and may belong to the same species, but the illustrations are too poor to be certain. *Christianella zhitangensis* (Liang, *in* Liu *et al.* 1983, pp. 277–278, pl. 95, figs 1–4) from the Huangnikang Formation (lower Ashgill) of Quxian, Zhejiang Province, East China, was also included within the synonymy of *Christiania* by Rong and Cocks (1994, p. 665) but can be specifically distinguished from our material in having a more transverse shell, a much stronger ventral median septum, and weaker dorsal socket ridges and anterior muscle bounding ridges. *C. nilssoni* (Sheehan 1973, pp. 65–66, pl. 3, figs 4–12), from the Jerrestad Mudstone (lower Ashgill) of southern Sweden and its contemporary rocks in East China, such as the Huangnikang Formation, differs from our material in having a more transverse shell with the width a little bigger than the length, a poorly impressed ventral muscle field without any bifurcating pallial markings, and a transverse S-shaped dorsal septum which originates from the posterior end of the inner side septa and extends antero-laterally to the middle of the outer side septa.

Superfamily CAMERELLOIDEA Hall and Clarke, 1894
Family PARASTROPHINIDAE Ulrich and Cooper, 1938

Genus EOSOTROPHINA Zhan and Rong, 1995

Eostrophina uniplicata (Liang, *in* Liu *et al.*, 1983)

Plate 7, figures 7, 11–15

1983 *Camerella uniplicata* Liang, *in* Liu *et al.*, p. 282, pl. 97, figs 1–6.
1995 *Eostrophina uniplicata* (Liang) Zhan and Rong, p. 569, pl. 4, figs 1–14; text-figs 11–13.

Remarks. Zhan and Rong (1995) reassigned Liang's *Camerella uniplicata*, from the Changwu Formation at Daqiao of Jiangshan County (Locality 6, Collection Jds), western Zhejiang, as the type species of a new genus *Eostrophina*. This genus was erected mainly on the basis of its well-

---

EXPLANATION OF PLATE 7

Figs 1–3. indeterminate teratelasmine; Changwu Formation, Dianbian, Locality 6, Horizon Jdd. 1, NIGP 128078; ventral internal mould; ×6. 2–3, NIGP 128079; internal and external moulds of dorsal valve; ×6.

Figs 4–6, 8–9. *Christiania* aff. *magna* Sheehan, 1987; Changwu Formation. 4, NIGP 128081; Locality 7, Horizon Jm 35; dorsal view of conjoined valves; ×3, 5, 9, NIGP 128082; Dianbian, Locality 6, Horizon Jdd; ventral and posterior views of ventral internal mould; ×3. 6, 8, NIGP 128083; Dianbian, Locality 6, Horizon Jdd; latex cast (6, ×4) and internal mould of dorsal valve (8, ×3).

Figs 7, 11–15. *Eostrophina uniplicata* (Liang, 1983); Changwu Formation, Shiyangwei, Locality 6, Horizon Jds. 7, 11–14, NIGP 124530; lateral, posterior, dorsal, ventral and anterior views of conjoined valves. 15, NIGP 124529; anterior view of conjoined valves. All ×2·5.

Fig. 10. *Foliomena folium* (Barrande, 1879); NIGP 128080; Changwu Formation, Locality 10, Horizon Js 2; ventral internal mould; ×10.

Figs 16–19. *Tcherskidium jiangshanensis* (Liang, 1983); Xiazhen Formation, Locality 1. 16–17, NIGP 128084; Horizon Fz 1; posterior and dorsal views of dorsal internal mould; ×3. 18–19, NIGP 128085; Horizon Yz 15; lateral and dorsal views of conjoined valves; ×2.

ZHAN and COCKS, late Ordovician brachiopods

developed and differentiated outer and inner brachial plates and the pair of short and fine alate plates in which it differs from all other parastrophinids. Some specimens associated with *Sowerbyella*, *Foliomena* and others, from the greyish-green mudstone of the upper part of the Pingliang Formation (lower Ashgill), Longxian of Shaanxi, North China, can also probably be identified as *Eostrophina* but are of a different species. These latter are the first record of the genus from outside the South China plate.

Superfamily PENTAMEROIDEA McCoy, 1844
Family VIRGIANIDAE Boucot and Amsden, 1963
Subfamily VIRGIANINAE Boucot and Amsden, 1963

Genus TCHERSKIDIUM Nikolaev and Sapelnikov, 1969

Type species. *Conchidium unicum* Nikolaev 1968, p. 47, pl. 2, figs 1–3; from the middle Ashgill of the Kolyma River, north-east Russia.

Remarks. Nikolaev and Sapelnikov (1969) differentiated *Tcherskidium* mainly on the basis of its very short outer brachial plates and its ribbing. It is close to *Eoconchidium* and *Proconchidium* in external morphology; however, *Eoconchidium* differs from it in having a ventral median septum much shorter than the spondylium, whilst *Proconchidium* has the ventral median septum longer than the spondylium and extending to the shell anterior margin. Because of the similarities in cardinalia and shell structure between *Tcherskidium* and *Virgiania*, Tcherskidiinae and Tcherskidiidae (established by Sapelnikov in 1972 and 1977 respectively) were later abandoned and *Tcherskidium* allocated (Boucot and Rong 1998) to the Virginaninae within the Virgianidae, which is where it was originally placed by Nikolaev and Sapelnikov (1969).

The relationship between the ventral median septum and the shell wall used to be thought to have familial significance, with the median septum wedging into the shell wall in the Pentameridae as opposed to continuous in the Virgianidae (Gauri and Boucot 1968). But both wedging and continuous relationships are seen in *Proconchidium* (Sapelnikov 1985) and the same occurs in *Tcherskidium*. So it is probable that this feature has no taxonomic significance at familial or even generic level. In addition, the anterior margin of the spondylium of *Tcherskidium* has some fine crenulations (seen in our serial sections; Text-fig. 11) similar to those of *Proconchidium* (Rong and Yang 1981). *Tcherskidium* is now known from the middle Ashgill of all equatorial palaeocontinents and therefore denotes no particular biogeographical ties between any two palaeoplates. However, the Chinese species described below is endemic.

EXPLANATION OF PLATE 8

Figs 1, 5. *Tcherskidium jiangshanensis* (Liang, 1983); Xiazhen Formation, Locality 1, Horizon Yz 15. 1, NIGP 128086; dorsal view of conjoined valves; ×2. 5, NIGP 128087; lateral view of conjoined valves; ×2.

Fig. 2. indeterminae brachiopod; NIGP 128088; Changwu Formation, Dianbian, Locality 6, Horizon Jdd; ventral internal mould; ×4.

Figs 3–4. indeterminate pentameroid; NIGP 128089; Changwu Formation, Locality 11, Horizon Chc 1; internal mould and latex cast of external mould of ventral valve; ×5.

Figs 6–14. *Altaethyrella zhejiangensis* (Wang, 1964); Xiazhen Formation, Locality 3, Horizon Ys. 6, 10, NIGP 128090; dorsal and anterior views of conjoined valves. 7–9, NIGP 128091; posterior, lateral and anterior views of conjoined valves. 11–14, NIGP 128092; posterior, dorsal, lateral and ventral views of conjoined valves. All ×2·5.

Figs 15–23. *Eospirigerina yulangensis* (Liang, 1983); Changwu Formation, Shiyangwei, Locality 6, Horizon Jds. 15–19, NIGP 128093; anterior, lateral, ventral, dorsal and posterior views of conjoined valves. 20–23, NIGP 128094; posterior, ventral, dorsal, and anterior views of conjoined valves. All ×2·5.

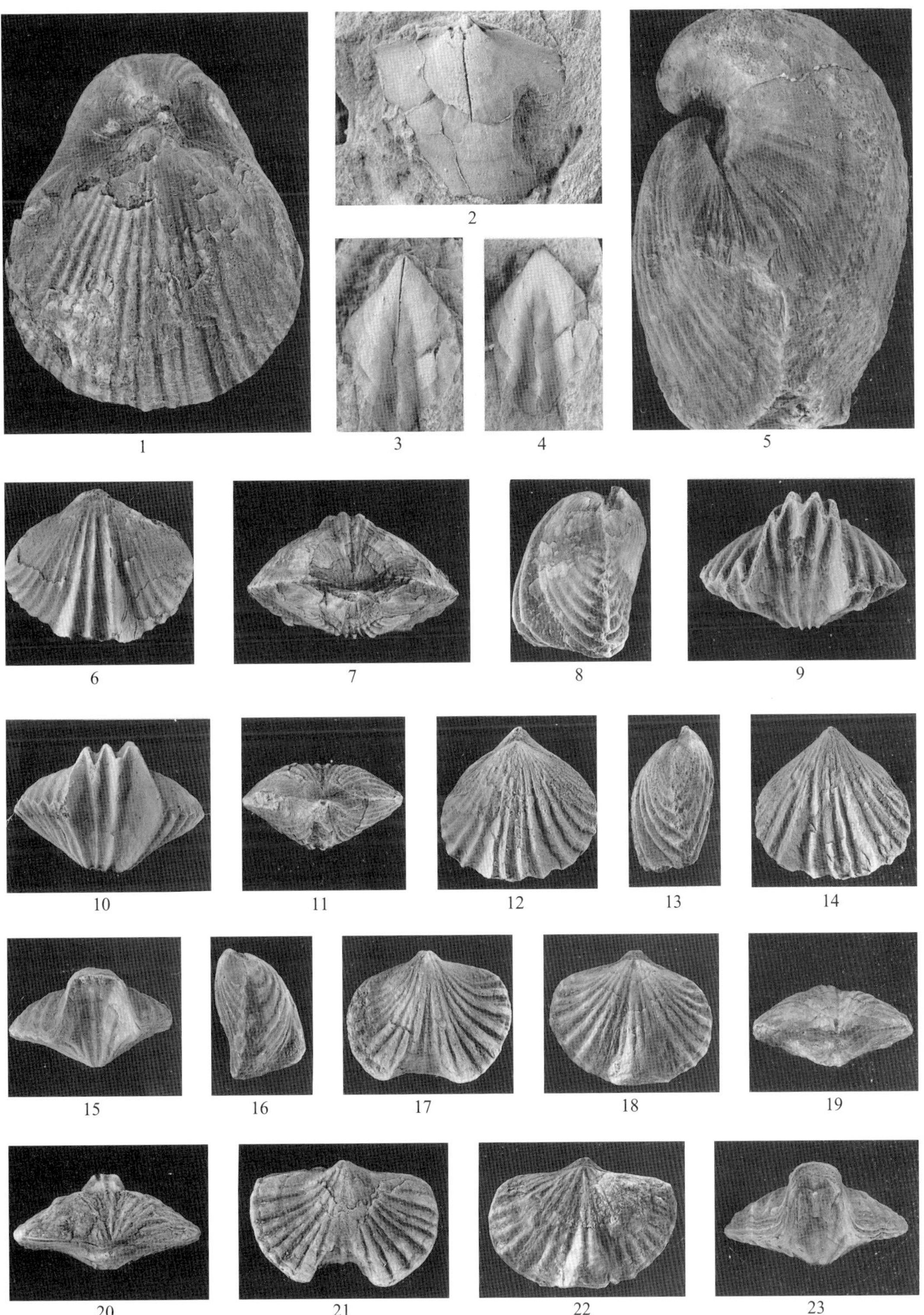

ZHAN and COCKS, late Ordovician brachiopods

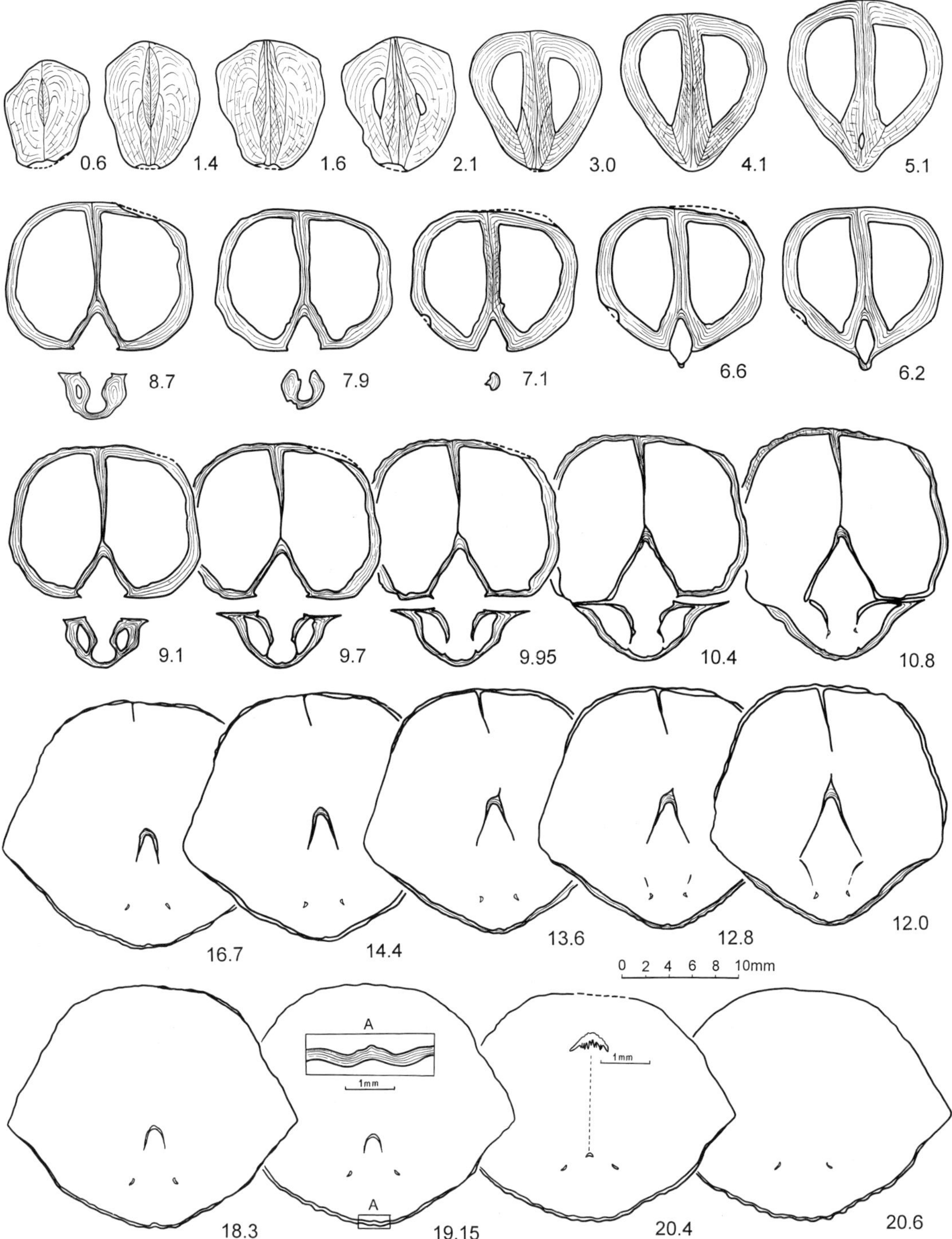

TEXT-FIG. 10. Transverse serial sections of *Tcherskidium jiangshanensis* (Liang, *in* Liu *et al.*, 1983) (43 sections made and 26 selected herein) from the Xiazhen Formation, Locality 1, Collection Yz 15. The enlargement at 19·15 mm is of the dorsal muscle field in the valve centre, and at 20·4 mm is of the fine crenulations of the anterior of the spondylium.

Tcherskidium jiangshanensis (Liang, *in* Liu *et al.*, 1983)

Plate 7, figures 16–19; Plate 8, figures 1, 5; Text-figure 10

1983 *Eoconchidium jiangshanensis* Liang, *in* Liu *et al.*, p. 283, pl. 95, figs 9–13.
1986 *Tcherskidium* sp. Rong and Han, p. 485, pl. 1, figs 1–3, 5, 7–8.

Material. Three thousand, three hundred and six body fossils amongst which 884 are conjoined valves, 944 are dorsal valves and 1478 are ventral valves, mainly from the lower and middle part of the Xiazhen Formation at Zhuzhai of Qunli, Yushan (Locality 1, Collections Yz 1 and Yz 15); 12 dorsal internal and seven external moulds, 55 ventral internal and 43 external moulds, mainly from the greyish-green mudstone of the Changwu Formation at Mulinlong of Tanshi, Jiangshian (Locality 7, Collections Jm 16, Jm-22 ~ 24).

Description: exterior. Shell 11·4–28·8 mm long, 12·7–24·3 mm wide with a length/width ratio of 0·9–1·2 (the largest individual 41·1 mm long, 29·2 mm wide and 25·1 mm thick). Outline elongated oval to nearly circular or rounded triangular; lateral profile ventri-biconvex, with the ventral beak strongly incurved. Maximum width at one-third or one-quarter of shell length. Weak dorsal fold and ventral sulcus variably developed. Wide palintrope with some transverse growth lines. Delthyrium and notothyrium open, with short and thin delthyrial and notothyrial ridges. Ornament of radial ribs, five or six per 5 mm near the shell anterior margin, branching in a few individuals at about one-third of the shell length, and growth lines on the anterior part of the shell.

Ventral interior. Teeth small. Wide and well-developed spondylium becoming narrower anteriorly, with some fine crenulations also anteriorly (Text-fig. 10), a little longer than the double median septum which is thinner anteriorly after its separation from the spondylium, extending to the shell midlength. Median septum wedges into the shell wall posteriorly (2·1 mm in Text-fig. 10) but is continuous with the shell wall anteriorly.

Dorsal interior. Cardinal process absent, inner socket ridges thin and low. Brachial apparatus of outer and inner brachial plates and brachial process. Outer brachial plates curved medially and subparallel to each other, less than 2 mm long (1·55 mm long in the sectioned individual which is medium sized; Text-fig. 10), and continuous with the shell wall. Much longer inner brachial plates extending antero-laterally to *c*. 17 per cent. of the shell length (4·3 mm long in the sectioned individual). Well-developed stick-like brachial processes posteriorly and plate-like anteriorly, free from the brachial plates for their anterior two-thirds.

Remarks. This is one of the most common brachiopods in the middle Ashgill of South China. Some variation can be observed: for example, among all the conjoined individuals nearly 60 per cent. have a rounded triangular outline with the length/width ratio about 1·3; 30 per cent. are slightly transverse with the ratio less than 1·2; and 10 per cent. are elongate with the ratio larger than 1·4. In addition nearly half of the specimens have a low dorsal fold, but only a few ventral valves have an apparent sulcus. *T. jiangshanensis* appears confined to South China. It differs from the type species, *T. unicum*, in having coarser ribs, a factor which appears even more significant because of the larger average size of *T. unicum*. There is no appreciable difference between the interiors of the two species and in the incurved beaks, and both species are very variable in length/width ratio. Most of our material is from the type locality of *Eoconchidium jiangshanensis* (Liang, *in* Liu *et al.* 1983, p. 283). Liang described a strong cardinal process in *E. jiangshanensis*, but this observation appears to be a mistake. The *Tcherskidium* sp. reported by Rong and Han (1986) is also from the same locality and horizon as the present species and is referred here to *T. jiangshanensis*.

indet. pentameroid?

Plate 8, figures 3–4

Remarks. A single ventral valve, with part and counterpart moulds, from the greyish-green mudstone of the basal part of the Changwu Formation at Huangnitang of Erduqiao, Changshan (Locality 11, Collection Chc 1), is 5·4 mm long and 3·8 mm wide. There is no ornament apart from densely populated growth lines, and there is a ventral sulcus within which there is a narrow median fold. There is a thin and short median septum up to 40 per cent. of the valve length. It is difficult to see beneath the umbo, but a short lateral slit on one side may represent the edge of a spondylium. Thus it is probably a pentamerid, but the family and genus are indeterminate. It is associated with *Skenidioides*, *Foliomena*, *Cyclospira* and small individuals of *Kassinella* (*Kassinella*), which together indicate a deeper-water environment.

Superfamily RHYNCHONELLOIDEA Gray, 1848
Family RHYNCHOTREMATIDAE Schuchert, 1913
Subfamily LEPIDOCYCLINAE Amsden, 1978

Genus ALTAETHYRELLA Severgina, 1978

Type species. *Altaethyrella megala* Severgina, 1978, p. 38, pl. 6, figs 7; from the upper Ordovician (Ashgill) of north-western Altai, Siberian Russia.

Altaethyrella zhejiangensis (Wang, 1964)

Plate 8, figures 6–14; Text-figure 11

1964 *Rhynchotrema zhejiangensis* Wang, *in* Wang and Jin, p. 46, pl. 12, figs 7–11.
1983 *Rhynchotrema gushanensis* Liang, *in* Liu *et al.*, p. 284, pl. 94, figs 6–10.
1997 *Altaethyrella zhejiangensis* (Wang); Zhan and Li, p. 435, pl. 1, figs 1–20, text-figs 2–6.

Remarks. Zhan and Li (1997) recently reviewed all the Ordovician species of *Rhynchotrema* reported from South and North China and found that they were all within the generic variability of *Altaethyrella*, a genus originally thought to be confined to Kazakhstan and the Altai Mountains area of Russia. They also discussed its taxonomic status, ecology and biogeography. *Altaethyrella* is extremely close to *Rhynchotrema* and *Rostricellula* in external morphology, but it never has the septalium in the dorsal interior which is always present in the latter two genera. *Rostricellula sarysuica* (Nikitin *et al.* 1996, pp. 92–93, figs 6N–S, 7, table 5), from an early Ashgill carbonate mound within the Dulankara Regional Stage of the northern Betpak-Dala Desert, central Kazakhstan, is nearly the same as *Altaethyrella zhejiangensis* in external morphology, including the

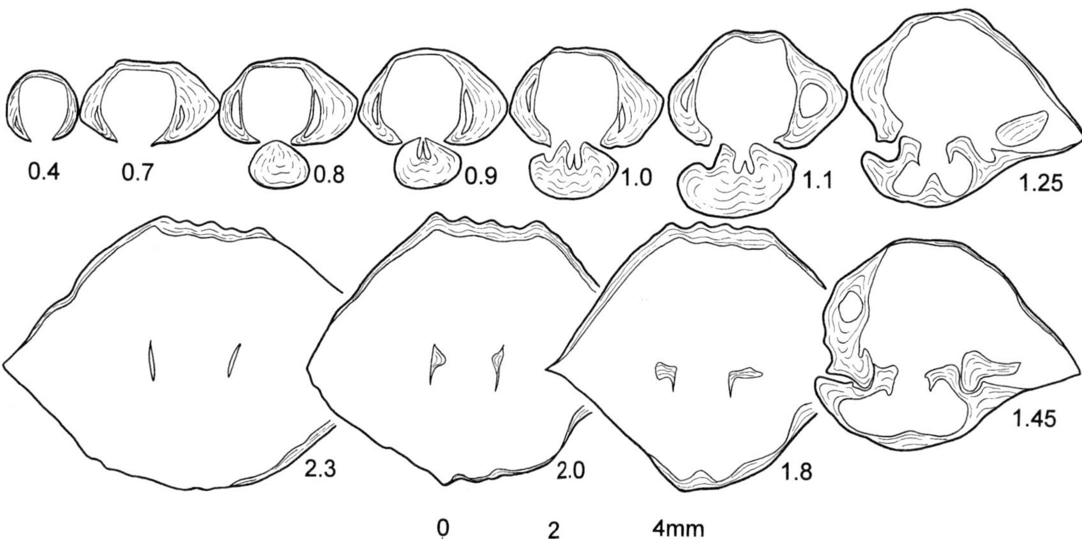

TEXT-FIG. 11. Transverse serial sections of *Altaethyrella zejiangensis* (Wang, *in* Wang and Jin, 1964) of conjoined valves with three plications in the sulcus (15 sections made and 11 selected herein); from the Xiazhen Formation, Locality 3, Collection Ys.

variable number of ribs on the dorsal fold and in the ventral sulcus; but *R. sarysuica* has a well-developed septalium supported by a high median septum which is absent in the interior of *A. zhejiangensis* (Text-fig. 11). True *Rostricellula* has not yet been found in China. *A. zhejiangensis* is one of the most common brachiopods in the Ashgill of our area, occurring at Localities 1 to 7, and most abundantly at Localities 2 and 3 (Locality 3 is Wang's type locality for this species). There is a large amount of variation within one population and between different populations of *A. zhejiangensis*: for example, the average size becomes larger north-eastward from Locality 1 to Locality 7; the shell outline ranges from transverse-rectangular to sub-spherical; the number of ribs in the sulcus varies from two to five; and the ventral beak varies from straight to strongly incurved.

The form identified as *Altaethyrella triplicata* (Fu) by Xu (1996, p. 555, pl. 3, figs 1–3; pl. 4, fig. 5; text-fig. 3), from the upper member of the Shiyanhe Formation (middle Ashgill) of Xichuan, Henan Province, has coarse, rounded and unbranching ribs, a weak fold and sulcus, a low dorsal median ridge and no cardinal process (according to Xu's serial sections) which are typical of *Ovalospira*. Fu's species *triplicata*, originally erected within *Drepanorhyncha* Cooper, 1956, has a triangular outline, strong fold and sulcus, a pair of well-developed dental plates and no dorsal median ridge, and is probably neither *Altaethyrella* nor *Ovalospira* and is certainly not *Drepanorhyncha*. *Latirhynchia inflata* Xu (1996, p. 556, pl. 4, figs 4–5; text-fig. 4), also from the upper member of the Shiyanhe Formation, has a strongly developed fold and sulcus, well-impressed ventral muscle field and a thin ridge-like cardinal process continuous with the low median ridge, all of which are typical of *Altaethyrella*. *A. inflata* (Xu, 1996) is very close to *A. zhejiangensis* but differs in having rounded ribs and no dental plates.

Superfamily ATRYPOIDEA Gill, 1871
Family ATRYPIDAE Gill, 1871
Subfamily SPIRIGERININAE d'Orbigny, 1849

Genus EOSPIRIGERINA Boucot and Johnson, 1967

1967 *Spirigerina* (*Eospirigerina*) Boucot and Johnson, p. 92.
1983 *Eorhynchula* Liang, *in* Liu *et al.*, p. 284.
1983 *Neorhynchula* Liang, *in* Liu *et al.*, p. 285.

Type species. *Zygospira putilla* Hall and Clarke, 1894, p. 365, pl. 54, figs 35–37; from the Hudson River Group (Ashgill), near Edgewood, Pike County, Missouri, USA.

Remarks. Boucot and Johnson (1967) named *Eospirigerina* as a subgenus within *Spirigerina*, with *Atrypa praemarginalis* as its type species. However, the type species was convincingly demonstrated by Amsden (1974) to be a junior synonym of *Zygospira putilla*. The subgenus was established mainly on the basis of its low ventral interarea, more incurved ventral beak, and poorly developed deltidial plates. Amsden (1974) determined that *E. putilla* from the Edgewood Group of Missouri had a unique feature: those individuals less than 8 mm long have a united hinge plate, but when the shell is 8–10 mm long, the hinge plates become separated but connected by a small arched plate. Mitchell (1977) thought that those differences between *Eospirigerina*, *Spirigerina* and *Plectatrypa* suggested by Boucot and Johnson (1967) were subjectively exaggerated, and probably variable within species. He also found that those specimens of *Eospirigerina* collected from the middle Ashgill of Pomeroy, Northern Ireland, had a well-impressed ventral muscle field, a feature originally taken as the main difference between *Eospirigerina* and *Plectatrypa*.

We agree with Copper (1982) in concluding that *Eospirigerina* differs from *Spirigerina* in having finer ribs, growth interruptions and overlapping growth lamellae in the anterior part of the shell, and internally thicker shells, better developed muscle scars and more massive teeth. *Eospirigerina*, with a range from Ashgill to upper Aeronian, is the earliest representative of the Atrypidae, whilst *Spirigerina* originated in the early Aeronian. We consider that the small specimens included in

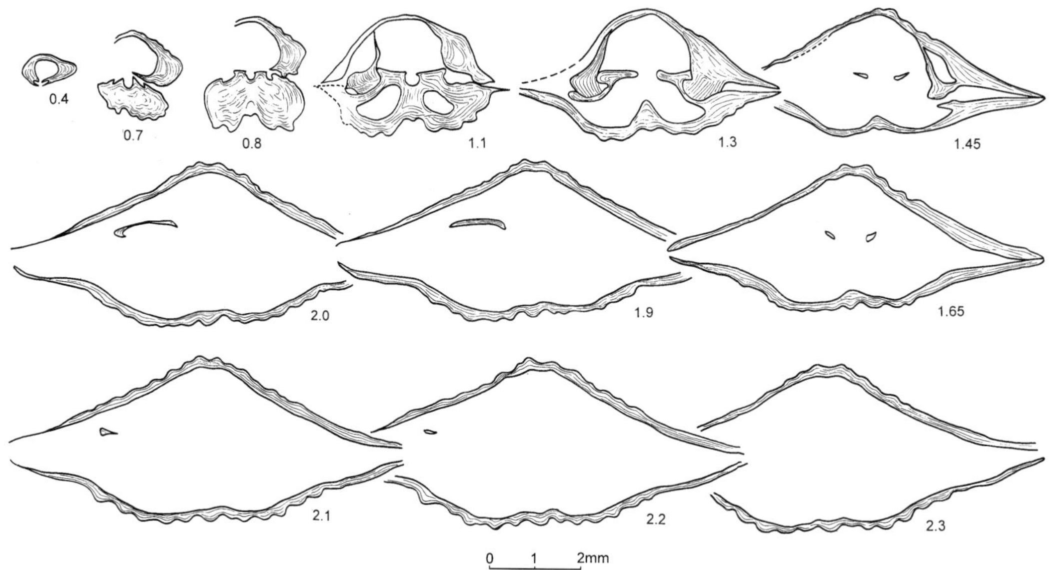

TEXT-FIG. 12. Transverse serial sections of *Eospirigerina yulangensis* (Liang, *in* Liu *et al.*, 1983) (21 sections made and 12 selected herein) from the Changwu Formation, Locality 6, Collection Jds.

E. putilla by Amsden (1974) with united hinge plates are probably representative of *Alispira* as revised by Jin and Norford (1992).

The two allegedly rhynchonellide genera *Eorhynchula* and *Neorhynchula*, erected by Liang (*in* Liu *et al.* 1983), have no similarity with rhynchonellides both externally and internally, and are identified here within *Eospirigerina* because of their fine branching ribs, straight ventral beak, strong fold and sulcus and well-impressed muscle field. The cardinal process described by Liang is not present.

Eospirigerina yulangensis (Liang, *in* Liu *et al.*, 1983)

Plate 8, figures 15–23; Text-figure 12

1983 *Eorhynchula yulangensis* Liang, *in* Liu *et al.*, p. 284, pl. 94, figs 11–15; pl. 98, fig. 9.
1983 *Neorhynchula jianglutangensis* Liang, *in* Liu *et al.*, p. 285, pl. 94, figs 1–5; pl. 98, fig. 8.

Material. Fifty-six conjoined valves, mainly from the greyish-green mudstone of the upper part of the Changwu Formation at Shiyangwei (300 m west of Shiyang) of Daqiao, Jiangshan (Locality 6, Collection Jds).

Description: exterior. Shells generally 7–10 mm long and 9–12·5 mm wide with length/width ratio 0·75–0·80. The largest shell is 10·5 mm long and 15·8 mm wide. Outline transverse to sub-rectangular, lateral profile convexo-planar or slightly convexo-concave; maximum width along the hinge line or slightly anterior of it. The latter is comparatively straight, but still non-strophic. Strongly convex dorsal valve thickest at two-thirds of the shell length. Dorsal fold and ventral sulcus extremely strong, whilst ventral valve has a very low median fold in the posterior one-third, continuous with the sulcus which is 30–40 per cent. of the shell width at the anterior margin. A fine dorsal median groove posteriorly continuous with the fold. Tongue long and high. Ventral beak always straight; simple interarea, with a pair of well-developed deltidial plates which sometimes join. Foramen epithyridid. Ornament mainly of radial ribs which branch occasionally on the two flanks, generally one to six ribs in the sulcus and on the fold, and 15–20 along the shell anterior margin. One or two growth interruptions and some irregular and discontinuous concentric lamellae often appear, particularly anteriorly.

Ventral interior. Teeth strong, thin and short dental plates close to the shell wall; small, well-impressed triangular muscle field at the posterior one-fifth to one-quarter of the shell.

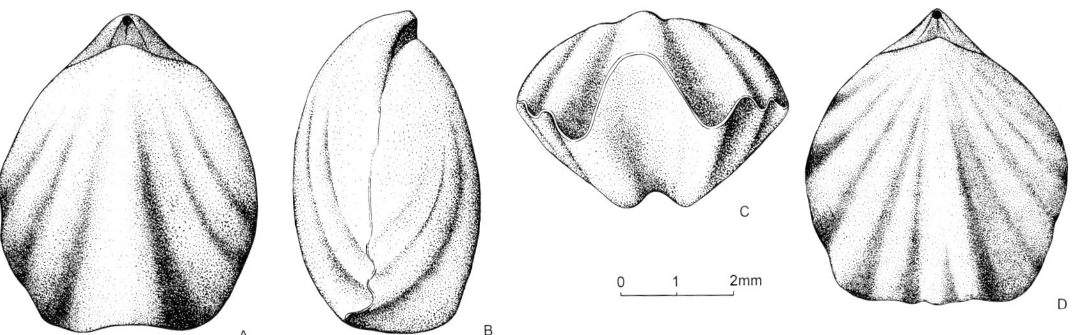

TEXT-FIG. 13. *Antizygospira liquanensis* Fu, 1982 from the Xiazhen Formation, Locality 1, Horizon Yz 17+. A–C, NIGP 128105; dorsal, lateral and anterior views of conjoined valves; D, NIGP 128106; dorsal view of another specimen from the same locality to illustrate the variation in ribbing styles.

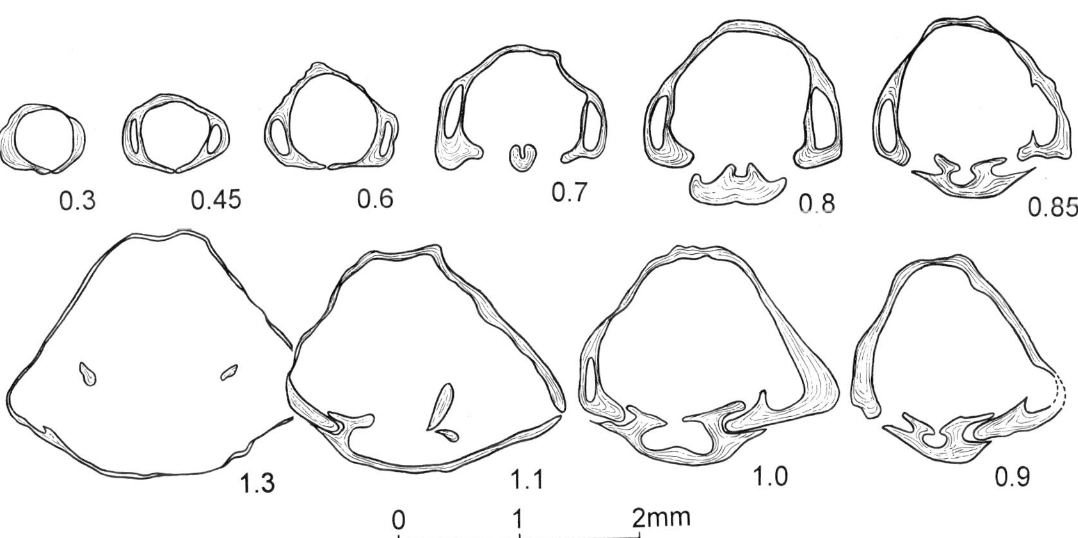

TEXT-FIG. 14. Transverse serial sections of *Antizygospira liqanensis* Fu, 1982 (14 sections made and ten selected herein) from the Xiazhen Formation, Locality 1, Collection Yz 17+.

Dorsal interior. Hinge plates well-developed, separated, extending subparallel; deep sockets with higher inner ridges; thick and short median ridge originating from the notohyrial cavity; crus thin and wide plate-like. No spiralia have been found (Text-fig. 12).

Measurements.	NIGP Cat. No.	L	W	L/W	T	W_2	W_3
	128093 (conjoined valves)	8·2	10·8	0·76	4·9	4·9	3·8
	128094 (conjoined valves)	8·1	12·3	0·66	5·5	5·5	3·6

Remarks. The type species of *Eorhynchula* and *Neorhynchula*, *E. yulangensis* and *E. jianglutangensis*, both occur in the Changwu Formation from Chun'an of Zhejiang (about 50 km north-east of our area). Unfortunately Liang (*in* Liu *et al.* 1983) illustrated only two specimens of each of his two new species *yulangensis* and *jianglutangensis*, all four of which are over twice the size of the specimens which we have found. These larger specimens have between seven and nine ribs in the fold and sulcus, whereas we have seen no more than six ribs in our smaller specimens. Thus there is an element of uncertainty in our attribution of our specimens to *yulangensis*; nevertheless, the presence of a very deep and high sulcus and fold in both Liang's specimens and our own, not seen consistently to such an extent in other species of *Eospirigerina* known to us, leads us to place our material within the first-named of Liang's species, *E. yulangensis*. *E. yulangensis* differs from *E. putilla* in its stronger and wider fold and sulcus, and a more transverse outline.

Spirigerina sinensis (Wang), from the mid Llandovery Xiangshuyuan Formation of Southwest China (Rong and Yang 1981, p. 225, pl. 17, figs 21–38, 40; pl. 22, fig. 13; text-figs 46–47), differs from our species in having a smaller shell, nearly biconvex profile, weak fold and sulcus, more ribs, and more concentric lamellae. Fu (1982) recognized three species of *Eospirigerina* from the late Ordovician Taoqupo Formation and its contemporary rocks in Shaanxi, North China: they are possibly conspecific with each other and require revision. *E. vetusta* Cocks and Modzalevskaya, 1997, from the Ashgill of Taimyr, Siberia, differs from *E. yulangensis* in its greater number of ribs, and weaker fold and sulcus.

Superfamily ZYGOSPIRIOIDEA Waagen, 1883
Family ZYGOSPIRIDAE Waagen, 1883
Subfamily ZYGOSPIRINAE Waagen, 1883

Genus ANTIZYGOSPIRA Fu, 1982

Type species. *Antizygospira liquanensis* Fu, 1982, from the Beiguoshan Formation (middle Ashgill), Dongzhuang of Liquan, Shaanxi, North China.

Remarks. *Antizygospira* is similar to *Zygospira*, but the former has a ventral sulcus and dorsal fold, whilst the latter has the reverse. *Antizygospira* is also very similar to *Alispira* externally, although the latter has a horizontal plate connecting the two hinge plates. No details about its spiralia and jugal processes are known. *Antizygospira* is confined to three areas: Shaanxi Province on the western margin of the North China Platform, the boundary area between Jiangxi and Zhejiang on the south-eastern part of Yangtze Platform, and Kazakhstan.

EXPLANATION OF PLATE 9

Figs 1–5. *Antizygospira liquanensis* Fu, 1982; Xiazhen Formation, Locality 1. 1–2, NIGP 128095; Horizon Yz 18; external and internal moulds of ventral valve; ×3. 3, NIGP 128096; Horizon Yz 7; dorsal internal mould; ×4. 4–5, NIGP 128097; Horizon Yz 17⁺; posterior and lateral views of conjoined valves; ×5.

Fig. 6. *Cyclospira* sp.; NIGP 128102; Changwu Formation, Locality 10, Horizon Js 2; ventral internal mould; ×8.

Figs 7–23. *Ovalospira dichotoma* Fu, 1982. 7–17, 21, Xiazhen Formation. 7–11, NIGP 128098; Locality 3, Horizon Ys; ventral, dorsal, lateral, posterior and anterior views of conjoined valves; ×2·5. 12, 14–17, NIGP 128099; Locality 1, Horizon Yz 4–3′; posterior, anterior, dorsal, ventral and lateral views of conjoined valves; ×2·5. 13, 21, NIGP 128100; Locality 1, Horizon Hz 4–3′; anterior and dorsal views of conjoined valves; ×2·5. 18–20, 22–23, NIGP 128101; Changwu Formation, Shiyangwei, Locality 6, Horizon Jds; anterior, dorsal, posterior, ventral and lateral views of conjoined valves; ×3.

Figs 24–27. *Eospirifer praecursor* Rong, Zhan and Han, 1994; Xiazhen Formation, Locality 1. 24, NIGP 124755; Horizon Yz 3–2; dorsal view of conjoined valves. 25, NIGP 124756; Horizon Yz 3–2; lateral view of conjoined valves. 26–27, NIGP 128103; Horizon Yz 11; external and internal moulds of dorsal valve with the posterior part of the ventral internal mould. All ×8.

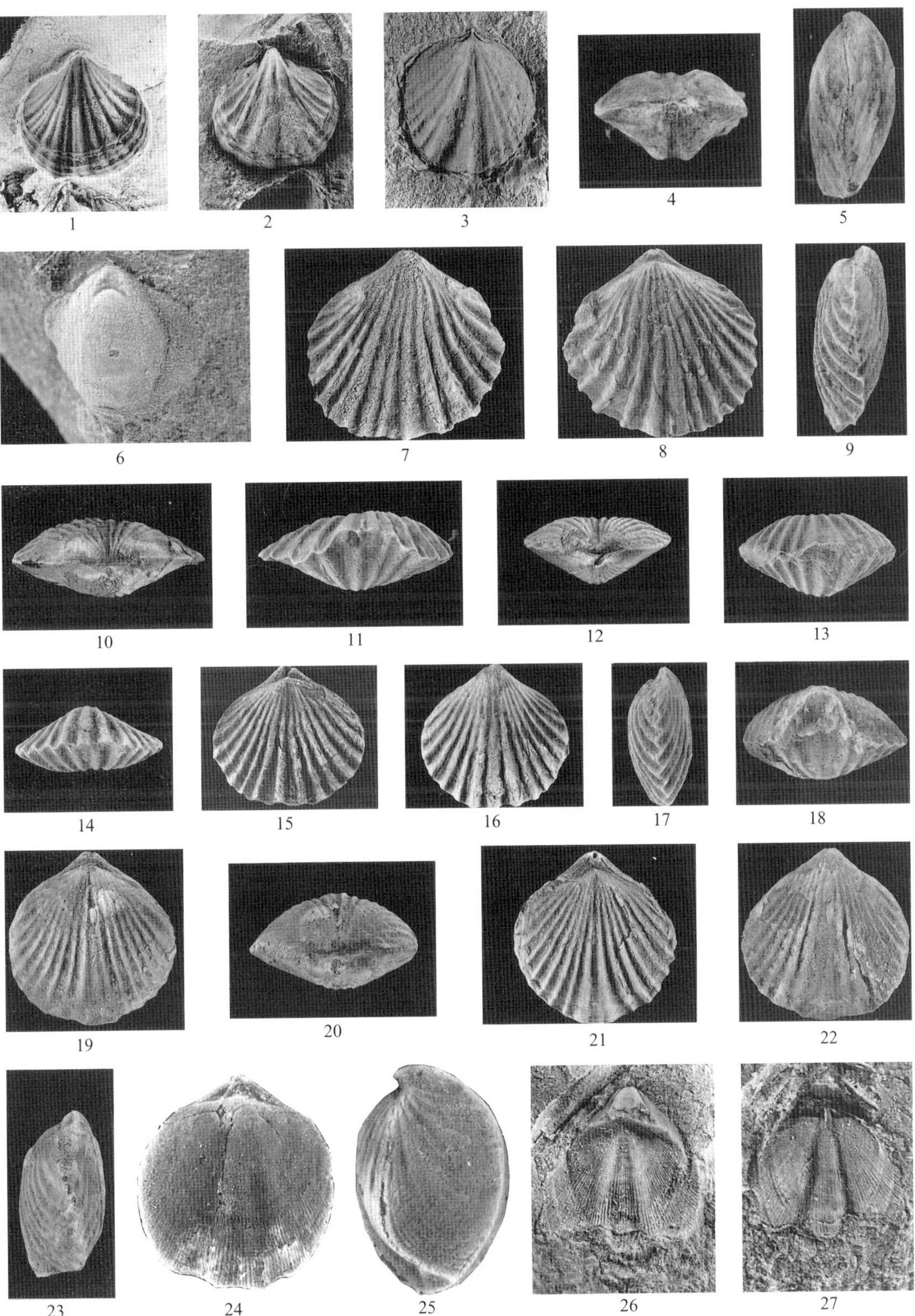

ZHAN and COCKS, late Ordovician brachiopods

Antizygospira liquanensis Fu, 1982

Plate 9, figures 1–5; Text-figures 13–14

1982 *Antizygospira liquanensis* Fu, p. 146, pl. 39, fig. 11, text-fig. 44.
1982 *Antizygospira dongzhuangensis* Fu, p. 146, pl. 39, fig. 12.
1982 *Antizygospira ornata* Fu, p. 146, pl. 39, fig. 13.
1986 *Zygospira* sp. Rong and Han, p. 486, pl. 1, fig. 6.

Material. Four hundred and sixty-eight conjoined valves and 1358 moulds; mainly from the middle to upper part of Xiazhen Formation at Zhuzhai and Tashan of Yushan (Localities 1 and 2, Collections Yz 7, Yz 17⁺, Yz 18, Yt 7 and Yt 9); and from the upper part of the Changwu Formation at Dianbian of Daqiao, Jiangshan (Locality 6, Collection Jdd).

Description: exterior. Shell 5·0–6·5 mm long and 4·5–5·5 mm wide with length/width ratio 1·15–1·30; the smaller specimens are more elongate. Outline longitudinally elliptical, maximum width two-thirds shell length. Lateral profile concavo-convex, dorsal valve variably concave or even slightly convex. A pair of strong ventral ribs marks each side of the sulcus; a shallow groove on the posterior part of dorsal valve, with a small rib in it, extends wider and higher rapidly to become the fold. Anterior commissure parasulcate, occasionally episulcate. Ventral beak straight or weakly curved; well-developed delthyrial plates abut; pedicle opening mesothyridid to epithyridid. Ornament mainly of radial ribs with occasional concentric lamellae; the ribs branching only once at about mid-length and varying greatly in number from five to 14 (Text-fig. 13).

Interior. Long and narrow teeth supported by thin and short dental plates; prominent dental cavity. Triangular muscle field well impressed. Small deep sockets with well developed inner socket ridges; strong discrete hinge plates extending medially and ventrally; short crus flat; only part of the first whorl of the spiralia preserved (Text-fig. 14).

Remarks. This is one of the most common elements of our fauna. Some ontogenetic variation can be observed. The smaller the shell, the more elongate the outline and the more convex the dorsal valve. Most specimens have only radial ribs, but a few growth interruptions and concentric lamellae are also seen in some individuals. The number of ribs appears to have nothing to do with shell size, but varies greatly in different individuals. The ribs on the fold or sulcus are generally absent, but some specimens (about one-third of the total) have one rib in the sulcus and two on the fold.

When establishing *Antizygospira*, Fu included three species: *A. dongzhuangensis* Fu, also from the type locality, differs in having a smaller shell and a more nearly plano-convex profile than *A. liquanensis*; and *A. ornata* Fu, from the upper part of the Beiguoshan Formation at Taoqupo of Yaoxian, Shaanxi, was distinguished in having a smooth ventral sulcus and only two ribs on each slope (Fu 1982). From the variation mentioned above, those characters used by Fu to differentiate the three species can be demonstrated in a single population of our material, and so Fu's three species are placed in synonymy here.

Zygospira (*Sulcatospira*) *minor* (Xu, *in* Jin *et al.*, 1979, p. 111, pl. 21, figs 1–5), from the late Ordovician Kemenzi Formation along the upper reaches of Tianbaohe River of Qilian, Qinhai, Northwest China, is similar to the present species externally but differs in having a weak ventral sulcus, and fewer ribs; however, we also assign it to *Antizygospira*. We also attribute *Zygospira parva* of Rukavishnikova (1956), from the Ashgill Dulankara Horizon of Kazakhstan, to *Antizygospira*, but it differs from *A. liquanensis* in having a weaker fold and sulcus, despite its larger size. *Alispira gracilis* Nikiforova, *in* Nikiforova and Andreeva, 1961 is also very similar to our species externally, but differs in having a horizontal plate connecting the hinge plates and capping the small cardinal cavity, whilst the hinge plates of *A. liquanensis* are always discrete (Text-fig. 14).

Genus OVALOSPIRA Fu, 1982

1982 *Ovalospira* Fu, p. 157.
1982 *Taoquopospira* Fu, p. 159.
1984 *Zygospira* (*Kuzgunia*) Misius, *in* Klenina *et al.*, p. 113.

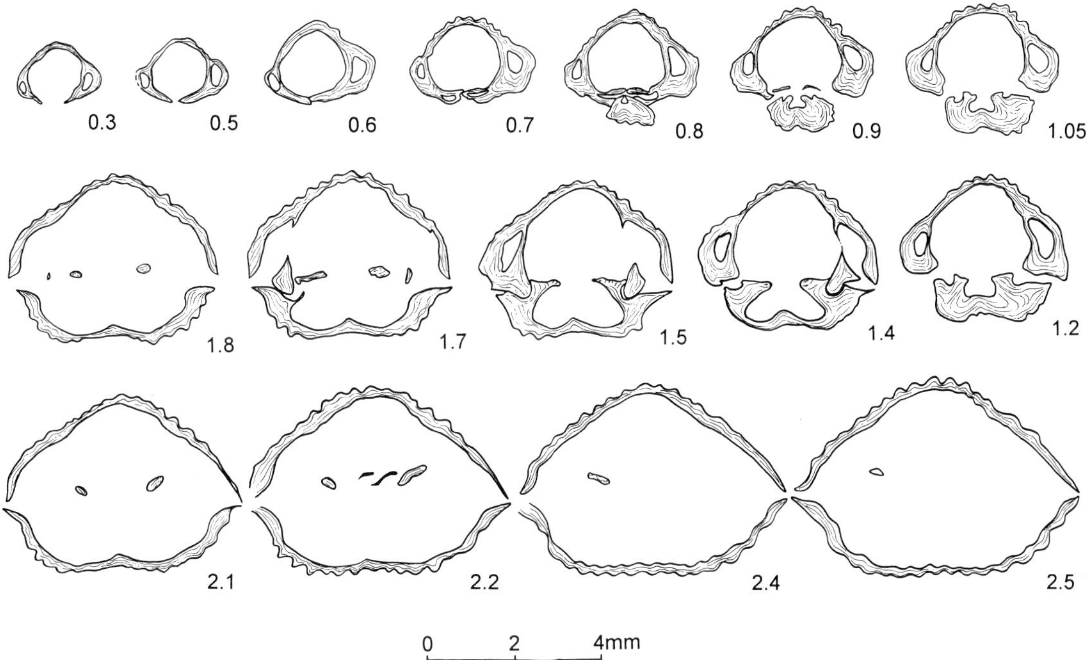

TEXT-FIG. 15. Transverse serial sections of *Ovalospira dichotoma* Fu, 1982 (23 sections made and 16 selected herein) from the Xiazhen Formation, Locality 1, Horizon Yz 4–3′.

Type species. Ovalospira ovalis Fu, 1982, from the Beiguoshan Formation (Ashgill), Ljiiapo of Longxian, Shaanxi, North China.

Remarks. The main characters of *Ovalospira* include: a very weak ventral sulcus and dorsal fold appearing in the anterior third of the shell; distinctive double-layered delthyrial plates which abut centrally, simple ribs on both flanks, sometimes branching mid-length; dental plates well-developed and hinge plates discrete (Text-fig. 15). These characters are also seen in *Taoquopospira* Fu, 1982, another endemic genus from the western margin of the North China Platform, and thus *Taoquopospira* is taken here as a junior synonym of *Ovalospira*. *Kuzgunia*, with its type species from the Dulankara Formation (Ashgill) of Kazakhstan, was erected as a subgenus of *Zygospira* by Misius (*in* Klenina *et al*. 1984) and also has the same internal and external characters as *Ovalospira*, which leads us to synonymize them.

Australispira Percival, from the upper Ordovician of New South Wales, Australia (Percival 1991, p. 169), can be distinguished from *Ovalospira* in having a dorsi-biconvex profile, stronger ventral sulcus and dorsal fold originating from the umbo, more ribs, poorly developed delthyrial plates, weak dental plates, and a ridge-like cardinal process between the separated hinge plates. *Alispira* Nikiforova, *in* Nikiforova and Andreeva, 1961 differs from *Ovalospira* in having a stronger fold and sulcus starting from the umbo, a ridge-like cardinal process, and a cardinal plate connecting the hinge plates. Fu (1982) erected *Orthocarina*, with its type species from the same type locality and horizon as *O. ovalis*, as a subgenus within *Ovalospira*. However, because its dorsi-biconvex shell, straight ventral beak, denser ribs, not completely fused delthyrial plates and submesothyridid foramen are all very distinct from *Ovalospira sensu stricto*, we consider *Orthocarina* to be a different genus.

The familial placing of *Ovalospira* is still uncertain. Copper (1990, p. 38) attributed it to the family Atrypidae, together with *Alispira*, *Eospirigerina*, *Spirigerina* and *Zygospiraella*. Because *Ovalospira* shares many similarities with *Zygospira*, and they are both a kind of primitive atrypoid with ribs, we provisionally allocate it to the Zygospiridae. *Ovalospira* is restricted to the upper Ordovician and is confined to South China, North China and Kazakhstan.

Ovalospira dichotoma Fu, 1982

Plate 9, figures 7–23; Text-figure 15

1982 *Ovalospira dichotoma* Fu, p. 157, pl. 41, fig. 5; text-fig. 63.

Material. Six hundred and nineteen conjoined valves and 210 moulds; mainly from the Xiazhen Formation at Zhuzhai, Shiyanshan and Tashan of Yushan (Localities 1–3, Collections Yz 4–3′, Yt 7 ~ 9, and Ys); and from the upper part of the Changwu Formation at Shiyangwei of Daqiao, Jiangshan (Locality 6, Collection Jds).

Description: exterior. Shell generally 8–12·0 mm long and 7·5–12·5 mm wide, with length/width ratio of 0·9–1·05; smaller individuals *c.* 6·0 mm long and 5·8 mm wide. Outline sub-circular to oval, lateral profile biconvex to ventri-biconvex. Maximum width at shell mid-length and maximum thickness posterior to mid-length. A shallow sulcus originating at two-thirds shell length and about one-third of shell width at the anterior margin. Anterior commissure weakly uniplicate. Ventral beak slightly incurved, interarea high, hinge line non-strophic, double-layered delthyrial plates uniting medially (see 0·6 mm and 0·7 mm in Text-fig. 15). Pedicle opening permesothyridid. Radial ribs becoming coarser with increasing shell length, but the number remaining stable, generally 16–20 along the anterior margin, seldom branching on the flanks, but more often on the fold and in the sulcus. One main central rib and a pair of weaker ones in the sulcus and three or four on the fold.

Ventral interior. Strong massive teeth; well-developed dental plates with clear dental cavities. Triangular muscle field most impressed in the centre.

Dorsal interior. Small notothyrial cavity with thick secondary shell deposition. Thick separate hinge plates, extending subparallel and slightly ventrally. Shallow sockets with well-developed inner bounding. Short crus plate-like anteriorly. Triangular muscle field with low wide myophragm (Text-fig. 15).

Remarks. Some population variation can be observed, such as the development of the sulcus and fold (from clearly developed to nearly absent), the shell outline from nearly circular to transversely oval, and the branching of ribs from twice to none. *O. parva* (Rukavishnikova), as revised by Klenina (*in* Klenina *et al.* 1984), from the Dulankara Formation (Ashgill) of Kazakhstan, is closest to *Ovalospira dichotoma* Fu, but the former has a larger shell, more ribs, smaller dental plates, and a much stronger dorsal median septum which is about one-third to one-half of the shell length, rather than the myophragm of *O. dichotoma*. The type species, *O. ovalis* Fu, can be distinguished from the present species by its larger shell and coarser ribs, with more rib branching in the sulcus or fold.

O. modesta (Misius, 1986, p. 195, pl. 24, figs 12–13), from the Lower Tezska series (middle Ordovician) of the northern slope of the Sary-Dzhaz range, Northern Kirgiz, is quite similar to *O. dichotoma* externally except for its narrower sulcus, but differs in having much longer dental plates and dorsal median ridge, and a short ridge-like cardinal process. *O. asiatica* Misius, 1986 (p. 187, pl. 22, figs 6–9; pl. 23, figs 1–5), from the same locality and horizon, and its three subspecies, *longula* (p. 190, pl 23, figs 10–13; pl. 24, figs 1–4; a little elongate), *lata* (p. 193, pl. 23, figs 6–9; a little less convex and more transverse) and *parva* (p. 191, pl. 24, figs 5–11; not the true *parva* as identified by Rukavishnikova and Klenina) are completely different from *O. dichotoma* externally and internally. They have a much stronger and narrower sulcus and fold which originate near the umbones of both valves, longer dental plates (the lateral bounding ridges of the ventral muscle field), and a short ridge-like cardinal process which is not present in any other species of *Ovalospira*. Thus *asiatica* and its three subspecies, which are probably better treated as a single taxon, need more substantial revision, and they will probably require assignment to a genus other than *Ovalospira*.

Superfamily LISSATRYPOIDEA Twenhofel, 1914
Family PROTOZYGIDAE Hall and Clarke, 1893
Subfamily CYCLOSPIRINAE Schuchert, 1913

Genus CYCLOSPIRA Hall and Clarke, 1893

Cyclospira sp.

Plate 9, figure 6

1988 *Cyclospira*? cf. *scanica* Sheehan; Cocks and Rong 1988, p. 55.

Material. Twelve ventral internal and five external moulds; mainly from the greyish-green mudstone of the Changwu Formation at Shangwu of Fengzu, Jiangshan (Locality 10, Collections Js 2).

Remarks. No dorsal interiors have been found and thus reliable specific identification cannot be made. However, the ventral interiors and the outline and profile of the exteriors lead to unquestionable attribution to *Cyclospira*. The identification is made easier by the consistent occurrence of this form with *Foliomena* in the distinctive deeper-water *Foliomena* assemblage, already recorded from the Changwu Formation by Cocks and Rong (1988). This led to the tentative identification with *C. scanica* Sheehan (1973) by Cocks and Rong, but since so many other species of *Cyclospira* are now known from rocks of Ashgill age in so many parts of the world, a particular specific identification would be inappropriate here for the rather poor Changwu Formation specimens.

C. jinheensis (Fu, 1982, p. 166, pl. 43, fig. 5), a late Ordovician species of *Cyclospira* from the Beiguoshan Formation at Dongzhuang of Liquan, Shaanxi, North China, is similar to our material externally including the shell size, but no interiors are known. *C. orbus* Cocks and Modzalevskaya, 1997 (pl. 5, figs 1, 3–6; text-figs 5–6), from the Korotkinskaya Formation (middle Ashgill) of the central Taimyr Peninsula, northern Siberia, Russia, differs from our material in having a much larger shell and a prominent ventral fold.

Superfamily CYRTIOIDEA Fredericks, 1924
Family CYRTIIDAE Fredericks, 1924
Subfamily EOSPIRIFERINAE Schuchert and LeVene, 1929

Genus EOSPIRIFER Schuchert, 1913

Eospirifer praecursor Rong, Zhan and Han, 1994

Plate 9, figures 24–27

1994 *Eospirifer praecursor* Rong, Zhan and Han, p. 772, figs 9.1–18.
1996*a* *Eospirifer praecursor* Rong and Zhan, p. 974, pl. 1, figs 1–17; pl. 3, figs 7, 16; pl. 4, fig. 14; text-figs 4–5, 27.

Remarks. Rong *et al*. (1994) described this species from moulds (Locality 9, Collection Jl 1) and Rong and Zhan (1996*a*) described the conjoined valves (Locality 1, Collections Yz 3–2 and Yz 15–1 ~ 9). It is the most widely distributed brachiopod in our area, with more than 5000 specimens recorded from localities 1–3 and 6–10, with lithologies ranging from argillaceous limestone to mudstone. According to Copper's (1973, 1977) studies, there are two kinds of connections between the spiralia in the atrypoids; one is a dorsally located, one-piece jugum (which is the more ancient), and the other is a more advanced development consisting of two discrete, ventrally located jugal processes which were probably connected by soft tissues when the animals were alive. The discovery of two jugal processes in *E. praecursor* indicates that the spiriferides probably possessed this feature from their origin.

indet. brachiopod

Plate 8, figure 2

Remarks. A unique internal mould was collected from the yellowish-green mudstone of the upper part of the Changwu Formation at Dianbian of Daqiao, Jiangshan (Locality 6, Collection Jdd). It has a small but well-developed spondylium or septalium which is undercut anteriorly and supported by a very thin and long median septum. There is doubt whether it is the ventral valve of a pentameroid or the dorsal valve of a rhynchonellide, or the dorsal valve of a dictyonellide. The beak would be extremely small for a pentameride and the proportions are unusual for a rhynchonellide. Some support for a dictyonellide identification lies in the faint impression of reticulate ornamentation which can be seen over the whole surface at high magnification, but which has not proved easy to photograph. With only a single valve interior, we prefer to leave this brachiopod unidentified.

Acknowledgements. We express our special thanks to Prof. Rong Jia-yu, Nanjing, for his great help, particularly in the initiation of the project, and to him and Dr David A. T. Harper, Galway, for valuable comments on the manuscript. We thank Prof. Fu Li-pu, Xi'an, for his help in the field, Deng Dong-xing and Phil Crabb for photography, and Ren Yu-gao for drawing some of the Text-figures. This work was supported by grants from the National Natural Science Foundation of China (Project No. 49372081) and the Special Supports for Palaeontology and Anthropology Research from Academia Sinica (Project No. 960405).

REFERENCES

AMSDEN, T. W. 1974. Late Ordovician and Early Silurian articulate brachiopods from Oklahoma, southwestern Illinois, and eastern Missouri. *Bulletin of the Oklahoma Geological Survey*, **119**, 1–154, pls 1–28.

—— 1978. Articulate brachiopods of the Quarry Mountain Formation (Silurian), eastern Oklahoma. *Bulletin of the Oklahoma Geological Survey*, **125**, 1–75, pls 1–13.

BAARLI, B. G. 1988. The Llandovery enteletacean brachiopods of the central Oslo region. *Palaeontology*, **31**, 1101–1129, pls 95–99.

BARRANDE, J. 1879. *Systême silurien du Centre de la Bohême. V.* Prague, Paris, 226 pp., 153 pls.

BIAN LI-ZENG, FANG YI-TING and HUANG ZHI-CHENG 1996. [On the types of Late Ordovician reefs and their characteristics in the neighbouring regions of Zhejiang and Jiangxi Provinces, South China.] 54–75, pls 1–2. *In* FAN JIA-SONG (ed.). [*The ancient organic reefs of China and their relations to oil and gas.*] Marine Publishing House, Beijing, 238 pp. [In Chinese].

BLAINVILLE, H. M. D. de 1824. *Dictionnaire des sciences naturelles*. 2nd edition. Paris, 324 pp.

BORISSIAK, M. A. 1956. Rod *Kassinella* [Genus *Kassinella*]. *Materiali vsesoyuznogo nauchno-issledovatelskii Geologischeskova Instituta, Moscow*, **12**, 50–52. [In Russian].

BOUCOT, A. J. 1983. Does evolution take place in an ecological vacuum? II. *Journal of Paleontology*, **57**, 1–30.

—— and AMSDEN, T. W. 1963. Virgianidae, a new family of pentameracean brachiopods. *Journal of Paleontology*, **37**, 296.

—— and JOHNSON, J. G. 1967. Silurian and Upper Ordovician atrypids of the genera *Plectatrypa* and *Spirigerina*. *Norsk Geologisk Tidsskrift*, **47**, 79–101, pls 1–4.

—— and RONG JIA-YU 1998. A global review of the earliest Pentameroidea (Brachiopoda: Ashgill, Late Ordovician). *Journal of Paleontology*.

CHEN XU, RONG JIA-YU, QIU JIN-YU, HAN NAI-REN, LI LUO-ZHAO and LI SHOU-JUN 1987. Late Ordovician stratigraphy, sedimentology and environment of Zhuzhai of Yushan, Jiangxi. *Journal of Stratigraphy*, **11**, 23–34. [In Chinese with English abstract].

—— —— WANG XIAO-FENG, WANG ZHI-HAO, ZHANG YUAN-DONG and ZHAN REN-BIN 1995. *Correlation of the Ordovician rocks of China (charts and explanatory notes)*. International Union of Geological Sciences Publication No. 31, 1–104.

CLOUD, P. E., Jr 1948. Brachiopods from the Lower Ordovician of Texas. *The Museum of Comparative Zoology, Harvard University*, **100**, 468–470, pl. 2.

COCKS, L. R. M. 1978. A review of British Lower Palaeozoic brachiopods, including a synoptic revision of Davidson's monograph. *Monograph of the Palaeontographical Society*, **131** (549), 1–256.

—— and FORTEY, R. A. 1988. Lower Palaeozoic facies and faunas around Gondwana. *Special Publication of the Geological Society*, **37**, 183–200.

—— —— 1997. A new *Hirnantia* fauna from Thailand and the biogeography of the latest Ordovician of south-east Asia. *Geobios*, **20**, 111–120, pls 1–2.

—— and LOCKLEY, M. G. 1981. Reassessment of the Ordovician brachiopods from the Budleigh Salterton Pebble Bed, Devon. *Bulletin of the British Museum (Natural History), Geology Series*, **35**, 111–124.

—— and MODZALEVSKAYA, T. L. 1997. Late Ordovician brachiopods from Taimyr, arctic Russia, and their palaeogeographical significance. *Palaeontology*, **40**, 1061–1093.

—— and RONG JIA-YU 1988. A review of the Late Ordovician *Foliomena* brachiopod fauna with new data from China, Wales and Poland. *Palaeontology*, **31**, 53–67.

—— —— 1989. Classification and review of the brachiopod superfamily Plectambonitacea. *Bulletin of the British Museum (Natural History), Geology Series*, **45**, 77–163.

COOPER, G. A. 1930. The brachiopod *Pionodema* and its homeomorphs. *Journal of Paleontology*, **4**, 369–382, pls 35–37.

—— 1956. Chazyan and related brachiopods. *Smithsonian Miscellaneous Collections*, **127**, 1–1245, pls 1–269.

—— and KINDLE, C. H, 1936. New brachiopods and trilobites from the Upper Ordovician of Percé, Quebec. *Journal of Paleontology*, **10**, 348–372, pls 51–53.

COPPER, P. 1973. The type species of *Lissatrypa* (Silurian Brachiopoda). *Journal of Paleontology*, **47**, 70–76, pls 1–2.

—— 1977. The late Silurian brachiopod genus *Atrypoidea*. *Geologiska Föreningens i Stockholm Förhandlingar*, **99**, 10–26.

—— 1982. Early Silurian atrypoids from Manitoulin Island and Bruce Peninsula, Ontario. *Journal of Paleontology*, **56**, 680–702, pls 1–3.

—— 1990. Evolution of the atrypid brachiopods. 35–40. *In* MacKINNON, D. and CAMPBELL, K. S. W. (eds). *Brachiopods through time*. Balkema, Rotterdam.

FOERSTE, A. F. 1924. Upper Ordovician faunas of Ontario and Quebec. *Memoir of the Canada Geological Survey*, **138**, 1–255, pls 10–15.

FORTEY, R. A. and COCKS, L. R. M. 1998. Biogeography and palaeogeography of the Sibumasu Terrane in the Ordovician: a review. *In* HALL, R. and ROSEN, B. R. (eds). *Biogeography and geological evolution of South-East Asia*. Balkema.

FREDERICKS, G. 1924. [On Upper Carboniferous spiriferids from the Urals.] *Geologich Komitet Izvestiya*, **38**, 295–324. [In Russian].

FU LI-PU 1982. Brachiopoda. 95–178, pls 30–45. *In* XI'AN INSTITUTE OF GEOLOGY AND MINERAL RESOURCES (ed.). [*Palaeontological atlas of Northwest China, Shaanxi-Gansu-Ningxia, Volume 1*.] Geological Publishing House, Bejiing, 480 pp., 106 pls. [In Chinese].

FÜRSICH, F. T. and HURST, J. M. 1974. Environmental factors determining the distribution of brachiopods. *Palaeontology*, **17**, 879–900.

GAURI, K. L. and BOUCOT, A. J. 1968. Shell structure and classification of Pentameracea McCoy, 1844. *Palaeontographica, Abteilung A*, **131**, 79–135, pls 6–23.

GILL, T. 1871. Arrangement of the families of molluscs prepared for the Smithsonian Institution. *Smithsonian Miscellaneous Collections*, **227**, 1–49.

GORYANSKY, V. Y. 1969. [Inarticulate brachiopods from the Cambrian and Ordovician deposits in the north-west Russian Platform.] *Materials on the Geology and Mineral Resources of the North-West Russian Platform*, **6**, 1–173, pls 1–21. [In Russian].

GRAY, J. E. 1848. On the arrangement of the Brachiopoda. *Annals and Magazine of Natural History*, **2**, 435–440.

HALL, J. 1847. Descriptions of the organic remains of the lower division of the New-York System. *Natural History of New York*, **1**, 1–338, pls 1–33.

—— 1859. Observations on genera of Brachiopoda. *12th Annual Report of New York State Cabinet*, 8–110.

—— and CLARKE, J. M. 1892–94. An introduction to the study of the genera of Palaeozoic Brachiopoda. *New York State Geological Survey, Palaeontology of New York*, **8**, (1), 1892, i–xvi, 1–367, pls 1–20; (2), 1893, i–xvi, 1–317; (3), 1894, 318–94, pls 21–84.

HARPER, D. A. T. 1984. Brachiopods from the upper Ardmillan succession (Ordovician) of the Girvan District, Scotland, Part 1. *Monograph of the Palaeontographical Society* **136** (564), 1–78, pls 1–11.

—— 1989. Brachiopods from the upper Ardmillan succession (Ordovician) of the Girvan District, Scotland, Part 2. *Monograph of the Palaeontological Society*, **142**, (579), 79–128, pls 12–22.

HAVLÍČEK, V. 1951. A paleontological study of the Devonian of Čelechovice; brachiopods (Pentameracea, Rhynchonellacea, Spiriferacea). *Sbornik Ústředního Ústavu Geologického*, **18**, 1–20, pls 1–4.

—— 1952. [On the Ordovician representatives of the family Plectambonitidae (Brachiopoda).] *Ústředního Ústavu Geologického*, **19**, 397–428, pls 1–3. [In Czech].

—— 1977. Brachiopods of the order Orthida in Czechoslovakia. *Ústředního Ústavu Geologického*, **44**, 1–327, pls 1–56.

—— and MERGL, M. 1982. Deep water shelly fauna in the latest Kralodvorian (Upper Ordovician, Bohemia). *Věstník Ústředního Ústavu Geologického*, **57**, 37–46.

—— and ŠTORCH, P. 1990. Silurian brachiopods and benthic communities in the Prague Basin (Czechoslovakia). *Věstník Ústředního Ústavu Geologického Svazek*, **48**, 1–275, pls 1–80.

JAANUSSON, V. and BASSETT, M. G. 1993. *Orthambonites* and related Ordovician brachiopod genera. *Palaeontology*, **36**, 21–63.

JIN JISUO and NORFORD, B. S. 1992. The Early Silurian atrypid brachiopod *Alispira* from Western Canada. *Palaeontology*, **35**, 775–800.

JIN YU-GAN, YE SONG-LING, XU HAN-KUI and SUN DONG-LI 1979. Brachiopoda. 60–217, pls 18–57. *In* NANJING INSTITUTE OF GEOLOGY AND PALAEONTOLOGY AND QINHAI INSTITUTE OF GEOLOGIC SCIENCES (ed.). [*Palaeontological atlas of Northwest China Qinhai, Volume 1.*] Geological Publishing House, Beijing, 387 pp., 132 pls. [In Chinese].

JONES, O. T. 1928. *Plectambonites* and some allied genera. *Memoirs of the Geological Survey of Great Britain, Palaeontology*, **1**, 367–527, pls 21–25.

KING, W. 1846. Remarks on certain genera belonging to the class Palliobranchiata. *Annals and Magazine of Natural History*, **18**, 26–42.

KLENINA, L. N., NIKITIN, I. F. and POPOV, L. E. 1984. [*Brachiopods and biostratigraphy of the Middle and Upper Ordovician of the Khingiz Mountains.*] Alma-Ata, 196 pp., 20 pls. [In Russian].

KOZŁOWSKI, R. 1929. Les brachiopodes gothlandiens de la Podolie Polonaise. *Palaeontologica Polonica*, **1**, 1–254, pls 1–12.

KULKOV, N. P. and SEVERGINA, L. G. 1989. [*Ordovician and Early Silurian stratigraphy and brachiopods from Gorny Altai.*] Nauka, Moscow, 221 pp., 32 pls. [In Russian].

LAI CAI-GENG, JIN RUO-GU, LIN BAO-YU and HUANG ZHI-GAO 1993. *Ordovician biofacies, sedimentological facies and palaeoenvironmental characteristics of Lower Yangtze region.* Geological Publishing House, Beijing, 87 pp. [In Chinese with English summary].

LAURIE, J. R. and BURRETT, C. 1992. Biogeographic significance of Ordovician brachiopods from Thailand and Malaysia. *Journal of Paleontology*, **66**, 16–23.

LENZ, A. C. 1977. Llandoverian and Wenlockian brachiopods from the Canadian Cordillera. *Canadian Journal of Earth Sciences*, **14**, 1521–1554.

LI LUO-ZHAO and HAN NAI-REN 1980. The discovery and its significance of the Ordovician trimerellid brachiopod fauna from western Zhejiang. *Acta Palaeontologica Sinica*, **19**, 8–22. [In Chinese with English abstract].

LIANG WEN-PING 1977. [*Stratigraphical review of Zhejiang Province, E China.*] Zhejiang Institute of Geologic Sciences, Hangzhou, 107 pp. [In Chinese].

LIN BAO-YU and ZOU XIN-GU 1977. Late Ordovician tabulate and heliolitoid corals from Zhejiang and Jiangxi Provinces and their stratigraphical significance. *Professional Paper of Stratigraphy and Palaeontology*, **3**, 108–208, pls 1–4. [In Chinese with English abstract].

—— —— 1986. Upper Ordovician rugose corals from Zheijang and Jiangxi Provinces. *Professional Paper of Stratigraphy and Palaeontology*, **14**, 1–22, pls 1–2. [In Chinese with English summary].

LIU, C. C. and CHAO, Y. T. 1927. Geology of southwestern Chekiang. *Bulletin of the Geological Survey of China*, **9**, 11–28.

LIU DI-YONG, XU HAN-KUI and LIANG WEN-PING 1983. Brachiopoda. 254–286, pls 88–89. *In* NANJING INSTITUTE OF GEOLOGY AND MINERAL RESOURCES (ed.). [*Palaeontological atlas of East China (1). Early Palaeozoic Volume*]. Geological Publishing House, Beijing, 657 pp., 176 pls. [In Chinese].

LU YAN-HAO, CHU CHAO-LING, CHIEN YI-YUAN, ZHOU ZHI-YI, CHEN JUN-YUAN, LIU GENG-WI, YU WEN, CHEN XU and XU HAN-KUI 1976. [Ordovician biostratigraphy and palaeozoogeography of China.] *Memoirs of the Nanjing Institute of Geology and Palaeontology*, **7**, 1–83. [In Chinese].

—— MU EN-ZHI, HOU YOU-TANG, ZHANG RI-DONG and LIU DI-YONG 1955. [New opinion on the Palaeozoic strata in west Zhejiang.] *Knowledge of Geology*, **2**, 1–6. [In Chinese].

McCOY, F. 1884. *A synopsis of the characters of the Carboniferous Limestone fossils of Ireland.* Dublin, 207 pp., 29 pls.

MENKE, C. T. 1828. *Synopsis methodica molluscorum generum omnium et speciarum earum quae in Museo Menkeano adservantur*. Pyrmont, 91 pp.

MITCHELL, W. I. 1977. The Ordovician Brachiopoda from Pomeroy, Co. Tyrone. *Monograph of the Palaeontographical Society*, **130** (545), 1–138, pls 1–28.

MISIUS, P. P. 1986. [*Ordovician brachiopods of northern Kirgizia*.] Frunze, 254 pp., 26 pls. [In Russian].

NEUMAN, R. B. and BRUTON, D. L. 1974. Early Middle Ordovician fossils from the Hølonda area, Trondheim region, Norway. *Norsk Geologisk Tidsskrift*, **54**, 69–115.

NIKIFOROVA, O. I. and ANDREEVA, O. N. 1961. [Stratigraphy of the Ordovician and Silurian of the Siberian Platform and its palaeontological significance (Brachiopods).] *Biostratigrafia Paleozoya Sibirskoi Platformy*, **1**, 1–142, pls 1–56. [In Russian].

NIKITIN, I. F. and POPOV, L. E. 1985. Ordoviskie Strofomenidy (Brakhiopody) Severnogo Priishim'ia (Tsentral'nyi Kazakhstan). *Ezhegodnik Vsesoiuznogo Paleontologischeskigo Obshchestva*, **28**, 34–49, pls 1–3. [In Russian].

—— —— 1996. Strophomenid and triplesiid brachiopods from an Upper Ordovician carbonate mound in central Kazakhstan. *Alcheringa*, **20**, 1–20.

—— —— and HOLMER, L. E. 1996. Late Ordovician brachiopod assemblage of Hiberno-Salairian type from Central Kazakhstan. *Geologiska Föreningens i Stockholm Förhandlingar*, **118**, 83–96.

—— —— and RUKAVISHNIKOVA, T. B. 1980. Brachiopoda. 37–73, pls 10–21. *In* APOLLONOV, M. K., BANDALETOV, S. M. and NIKITIN, I. F. (eds). *Granitsa Ordovika i Silura v Kazakhstane*. Nauka KazSSR, Alma-Ata, 300 pp., 56 pls. [In Russian].

NIKOLAEV, A. A. 1968. *Conchidium*(?) *unicum* sp. nov. 47–48. *In* BALASHOV, Z. G. (ed.). [*Field atlas of the Ordovician fauna of Northeast USSR*.] Magadan, 147 pp. [In Russian].

—— and SAPELNIKOV, V. P. 1969. [Two new genera of Late Ordovician Virgianidae.] *Trudy Sverdlovskogo Ordena Trudovogo Drasnogo Znameni Gornogo Instituta*, **63**, 11–17. [In Russian].

ÖPIK, A. A. 1930. Brachiopoda Protremata der Estländischen Ordovizischen Kukruse-Stufe. *Acta et Commentationes Universitatis Tartuensis*, **17**, 1–262, pls 1–22.

ORBIGNY, A. d'1848–1851. *Paléontologie française, Terrains Crétacés, 4, Brachiopoda*. Paris, 1–390, pls 490–599.

PERCIVAL, I. G. 1978. Inarticulate brachiopods from the Late Ordovician of New South Wales, and their palaeoecological significance. *Alcheringa*, **2**, 117–141.

—— 1979*a*. Ordovician plectambonitacean brachiopods from New South Wales. *Alcheringa*, **3**, 91–116.

—— 1979*b*. Late Ordovician articulate brachiopods from Gunningbland, central western New South Wales. *Proceedings of Linnean Society of New South Wales*, **103**, 175–187.

—— 1991. Late Ordovician articulate brachiopods from central New South Wales. *Memoirs of the Association of Australasian Palaeontologists*, **11**, 107–177.

POPE, J. K. 1976. Comparative morphology and shell histology of the Ordovician Strophomenacea (Brachiopoda). *Palaeontographica Americana*, **49**, 127–213.

POPOV, L. E. 1980. [New species of brachiopod from the Middle Ordovician of the Chu-Ili Hills.] *Annual of the All-Union Paleontological Society*, **23**, 139–158. [In Russian].

POTTER, A. W. 1990. Middle and Late Ordovician brachiopods from the eastern Klamath Mountains, northern California, part 2. *Palaeontographica, Abteilung A*, **213**, 1–114, pls 1–10.

QIAN JIA-JU 1987. Late Ordovician (Wufengian) deposits and the boundary between Ordovician and Silurian in western Zhejiang, E China. *Acta Geologica Sinica*, **2**, 101–112. [In Chinese with English abstract].

RONG JIA-YU 1984*a*. Brachiopods of latest Ordovician in the Yichang district, western Hubei, Central China. 111–190, pls 1–14. *In* NANJING INSTITUTE OF GEOLOGY AND PALAEONTOLOGY, ACADEMIA SINICA (ed.). *Stratigraphy and palaeontology of systemic boundaries in China. Ordovician-Silurian Boundary*, (*1*). Anhui Science and Technology Publishing House, Hefei, 516 pp.

—— 1984*b*. [Ecostratigraphic evidence of the Upper Ordovician regressive sequence and the effect of the glaciation.] *Journal of Stratigraphy*, **8**, 19–29. [In Chinese].

—— and CHEN XU 1987. Faunal differentiation, biofacies and lithofacies pattern of Late Ordovician (Ashgillian) in South China. *Acta Palaeontologica Sinica*, **26**, 507–535. [In Chinese with English summary].

—— and COCKS, L. R. M. 1994. True *Strophomena* and a revision of the classification and evolution of strophomenid and 'strophodontoid' brachiopods. *Palaeontology*, **37**, 651–694.

—— and HAN NAI-REN 1986. Preliminary report on Upper Ordovician (mid-Ashgillian) brachiopods from Yushan, northeastern Jiangxi (Eastern China). *Biostratigraphie du Paléozoique*, **4**, 485–490, pl. 1.

—— and HARPER, D. A. T. 1988. A global synthesis of the latest Ordovician Hirnantian brachiopod faunas. *Transactions of the Royal Society of Edinburgh: Earth Sciences*, **79**, 383–402.

—— —— ZHAN REN-BIN and LI RONG-YU 1994. *Kassinella-Christiania* Associations in the early Ashgill *Foliomena* brachiopod fauna of South China. *Lethaia*, **27**, 19–28.

—— LI RONG-YU and KULKOV, N. P. 1995. Biogeographic analysis of Llandovery brachiopods from Asia with a recommendation of use of affinity indices. *Acta Palaeontologica Sinica*, **34**, 428–453.

—— and YANG XUE-CHANG 1981. Middle and late Early Silurian brachiopod faunas in Southwest China. *Memoirs of the Nanjing Institute of Geology and Palaeontology, Academia Sinica*, **13**, 163–270, pls 1–26. [In Chinese with English summary].

—— and ZHAN REN-BIN 1996*a*. Brachidia of Late Ordovician and Silurian eospiriferines (Brachiopoda) and the origin of the spiriferides. *Palaeontology*, **39**, 941–977.

—— —— 1996*b*. Distribution and ecological evolution of the *Foliomena* fauna (Late Ordovician brachiopods). 90–97, pls 1–2. *In* WANG HONG-ZHEN and WANG XUN-LIAN (eds). *Centennial Memorial Volume of Prof. Sun Yunzhu: palaeontology and stratigraphy*. China University of Geosciences Press, Beijing, 327 pp.

—— —— and HAN NAI-REN 1994. The oldest known *Eospirifer* (Brachiopoda) in the Changwu Formation (Late Ordovician) of western Zhejiang, East China, with a review of the earliest spiriferoids. *Journal of Paleontology*, **68**, 763–776.

ROUAULT, M. 1850. Note préliminaire sur une nouvelle formation découverte dans le terrain silurien inférieur de la Bretagne. *Bulletin de la Société Géologie de France*, **7**, 724–744.

RUKAVISHNIKOVA, T. B. 1956. [Ordovician brachiopods from Kazakhstan.] *Trudy Akademiia NAUK SSSR, Geologicheskii Institut*, **1**, 105–168, pls 1–5. [In Russian].

SALMON, E. S. 1942. Mohawkian Rafinesquininae. *Journal of Paleontology*, **16**, 564–603, pls 85–86.

SAPELNIKOV, V. P. 1972. Siluriiskie Pentameracea vostochnogo sklona Srednego i Severnogo Urala. *Instituta Geologii i Geokhimii Uralskii Nauchnyi Tsentr, Akademiya Nauk SSSR*, 1–295. [In Russian].

—— 1977. [System and phylogeny of superfamily Stricklandiacea (Brachiopoda).] *Instituta Geologii i Geokhimii, Uralskii Nauchnyi Tsentr, Akademiya SSSR*, **129**, 3–19. [In Russian].

—— 1985. [System and stratigraphic significance of the suborder Pentameridina brachiopods.] *Instituta Geologii i Geokhimii, Ural'skii Nauchniyi Tsentr, Akademiya Nauk SSSR*, 1–205, pls 1–48. [In Russian].

SCHUCHERT, C. 1913. Class 2. Brachiopoda. 355–420. *In* ZITTEL, K. A. von (ed.). *Textbook of Palaeontology*, 1. 2nd edition. London.

—— and COOPER, G. A. 1931. Synopsis of the brachiopod genera of the suborders Orthoidea and Pentameroida, with notes on the Telotremata. *American Journal of Science*, **22**, 241–251.

—— and LEVENE, C. M. 1929. Brachiopoda (Generum et genotyporum index et bibliographia). *Fossilium Catalogus*, 1 Animalia (42), Berlin, 1–140.

SEVERGINA, L. G. 1978. Brakhiopody i stratigrafiya verkhnego ordovica Gornovo Altaya, Salaira i Gornoy Shorii. *Trudy Instituta Geologii i Geofiziki, Sibirskoye Otdeleniye, Akademiya Nauk SSSR*, **405**, 3–41. [In Russian].

SHEEHAN, P. M. 1973. Brachiopods from the Jerrestad Mudstone (Early Ashgillian, Ordovician) from a boring in southern Sweden. *Geologica et Palaeontologica*, **7**, 59–76, pls 1–3.

—— 1987. Late Ordovician (Ashgillian) brachiopods from the region of the Sambre and Meuse Rivers, Belgium. *Bulletin de l'Institut Royal des Sciences Naturelles de Belgique, Sciences de la Terre*, **57**, 5–81, pls 1–16.

—— and COOROUGH, P. J. 1990. Brachiopod zoogeography across the Ordovician-Silurian extinction event. *Memoir of the Geological Society, London*, **12**, 181–187.

—— and LESPÉRANCE, P. J. 1979. Late Ordovician (Ashgillian) brachiopods from the Percé region of Quebec. *Journal of Paleontology*, **53**, 950–967.

SHENG, S. F. 1951. [Stratigraphy of Zhejiang Province.] *Zhejiang Geology*, **2**, 1–18. [In Chinese].

SINCLAIR, G. W. 1945. Some Ordovician lingulid brachiopods. *Transactions of the Royal Society of Canada*, **39**, 55–82, pls 1–4.

TWENHOFEL, W. H. 1914. The Anticosti Island faunas. *Bulletin of Canada Geologic Survey, Geology Series*, **19**, 1–39, pl. 1.

ULRICH, E. O. and COOPER, G. A. 1936. New genera and species of Ozarkian and Canadian brachiopods. *Journal of Paleontology*, **10**, 616–631.

—— —— 1938. Ozarkian and Canadian Brachiopoda. *Special Paper of the Geological Society of America*, **13**, 1–323, pls 1–58.

—— —— 1942. New genera of Ordovician brachiopods. *Journal of Paleontology*, **16**, 620–626, pl. 90.

VERNEUIL, E. de 1848. Note sur quelques brachiopodes de L'ile de Gothland. *Bulletin of the Geological Society of France*, **2**, 339–347.

WAAGEN, W. H. 1883. Salt Range fossils, Part 4 (2), Brachiopoda. *Memoirs of the Geological Survey of India, Palaeontologica Indica*, (13) **1**, 391–546, pls 29–49.

WALMSLEY, V. G. and BOUCOT, A. 1971. The Resserellinae – a new subfamily of Late Ordovician to Early Devonian dalmanellid brachiopods. *Palaeontology*, **14**, 487–531, pls 91–102.

WANG XIAO-FENG, NI SHI-ZHAO, ZENG QING-LUAN, XU GUANG-HONT, ZHOU TIAN-MEI, LI ZHI-HONG, XIANG LI-WEN and LAI CAI-GENG 1987. *Biostratigraphy of the Yangtze Gorge area 2: Early Palaeozoic Era.* Geological Publishing House, Beijing, 641 pp., 72 pls. [In Chinese with English summary].

WANG YU 1956. New species of Brachiopoda I. *Acta Palaeontologica Sinica*, **4**, 1–24, pls 1–5.

—— and JIN YU-GAN 1964. Brachiopods. 46, pl. 12. *In* NANJING INSTITUTE OF GEOLOGY AND PALAEONTOLOGY, ACADEMIA SINICA (ed.). [*A handbook of standard fossils from South China.*] Science Press, Beijing, 233 pp. [In Chinese].

WILLIAMS, A. 1953. The classification of the strophomenoid brachiopods. *Journal of Washington Academic Sciences*, **43**, 1–13.

—— 1962. The Barr and Lower Ardmillan Series (Caradoc) of the Girvan District, south-west Ayrshire, with descriptions of the Brachiopoda. *Memoir of the Geological Society, London*, **3**, 1–267, pls 1–25.

—— *et al.* 1965. *Treatise on invertebrate paleontology. Part H. Brachiopoda.* University of Kansas Press, Kansas, 927 pp.

WILSON, A. E. 1913. A new brachiopod from the base of the Utica. *Bulletin of the Canadian Geological Survey, Victoria Memorial Museum*, **1**, 81–86, pl. 8.

WOODWARD, S. P. 1852. *A manual of the Mollusca; or rudimentary treatise of recent and fossil shells.* London, xvi + 1–486, pls 1–24.

WRIGHT, A. D. 1964. The fauna of the Portrane Limestone, II. *Bulletin of the British Museum (Natural History), Geology Series*, **9**, 157–256, pls 1–11.

—— 1968. A new genus of dicoelosiid brachiopod from Dalarna. *Arkiv för Zoologi*, **2**, 127–138, pl. 1.

—— 1993. Subdivision of the Lower Palaeozoic articulate brachiopod family Triplesiidae. *Palaeontology*, **36**, 481–493.

XU HAN-KUI 1996. Late Ordovician brachiopods from central part of eastern Qinling region. *Acta Palaeontologica Sinica*, **35**, 544–574, pls 1–6. [In Chinese with English summary].

—— and LIU DI-YONG 1984. Late Early Ordovician brachiopods of Southwest China. *Memoirs of Nanjing Institute of Geology and Palaeontology, Academia Sinica*, **8**, 147–235, pls 1–18. [In Chinese with English summary].

YANG, S. 1984. Ordovician tabulate coral assemblages of Britain and their zoogeographical relationships. *Palaeontographica Americana*, **54**, 169–173.

YU JIAN-HUA, BIAN LI-ZENG, HUANG ZHI-CHENG, CHEN MEI-JUAN, FANG YI-TING, ZHOU XIAO-PING and SHI GUI-JUN 1992. [Preliminary research on the Late Ordovician reefs from the border region of Zhejiang and Jiangxi Provinces, South China.] *Journal of Nanjing University (Earth Sciences)*, **4**, 1–12. [In Chinese].

ZHAN REN-BIN and FU LI-PU 1994. New observations on the Upper Ordovician stratigraphy of Zhejiang-Jiangxi border region, E China. *Journal of Stratigraphy*, **18**, 267–274. [In Chinese with English abstract].

—— and LI RONG-YU 1997. The discovery of *Altaethyrella* Severgina 1978 in China (Late Ordovician rhynchonelloid brachiopod). *Acta Palaeontologica Sinica*, **36**, 429–441, pl. 1.

—— and RONG JIA-YU 1994. *Tashanomena*, a new strophomenoid genus from the middle Ashgill rocks (Ordovician) of Xiazhen, Yushan, NE Jiangxi, East China. *Acta Palaeontologica Sinica*, **33**, 416–428, pl. 1.

—— —— 1995. Four new Late Ordovician brachiopod genera from the Zhejiang-Jiangxi border region, East China. *Acta Palaeontologica Sinica*, **34**, 549–574, pls 1–4.

ZHANG NING and BOUCOT, A. J. 1988. *Epitomyonia* (Brachiopoda): ecology and functional morphology. *Journal of Paleontology*, **62**, 753–758.

ZHAN REN-BIN

Nanjing Institute of Geology and Palaeontology
Academia Sinica, Chi-Ming Ssu
Nanjing 210008
The People's Republic of China

L. R. M. COCKS

Department of Palaeontology
The Natural History Museum
Cromwell Road, London SW7 5BD, UK

Typescript received 28 April 1997
Revised typescript received 27 August 1997

APPENDIX

Faunal lists used in the calculation of affinity indices; the identifications are as given by the quoted authors, apart from taxa with an asterisk (*) which we have modified:

1. Middle part of Beiguoshan Formation (middle Ashgill), Shaanxi Province, North China (Fu 1982); *Pionorthis beiguoshanensis*; *Austinella kankakensis*; *Schizophorella shaanxiensis*; *Platystrophia multiplicata*; *Bicuspina trapezoida*; *B. planodorsa*; *Sowerbyella fupingensis*; *Kjerulfina polycyma*; *Christiania gigantea*; *Parastrophinella minor*; *Drepanorhyncha pentagonia*; *Latirhynchia transversa*; **Altaethyrella yaoxianensis*; *Zygospira shaanxiensis*; *Z. rhomboidalis*; **Antizyospira liquanensis*; **Eospirigerina yaoxianensis*; **Ovalospira dichotoma*; **Orthocarina carinatiformis*; *Idiospira minor*; *I. taoqupoensis*; *Cyclospira jingheensis.*

2. Upper part of Chokpar Formation and Dulankara Horizon (middle Ashgill), Dulankara, Kazakhstan (Nikitin *et al.* 1980; Klenina *et al.* 1984); *Plaesiomys* (*Dinorthis*) *abayi*; *P.* (*D.*) *tschinghisensis*; *P.* (*D.*) *nikitini*; *P.* (*D.*) *distincta*; *Mimella brevis*; **M. zhejiangensis recta*; *Schizophorella* aff. *kasachstanica*; *Dalmanella pulchra*; *Paucicrura tschinghisensis*; *Triplesia subcraigensis*; *Leptellina* (*Mabella*) *semiovalis*; *L.* (*M.*) *obtusa*; *L.* (*M.*) *incurvata*; *Sowerbyella* (*Sowerybeylla*) *papiliuncula*; *S.* (*S.*) *akdombakensis*; *S.* (*Rugosowerbyella*) cf. *ambigua*; *S.* (*Viruella*) *insueta*; *Eoplectodonta*? sp.; *Kassinella tschinghisensis*; *K. nana*; *K.* sp.; *Strophomena phyaliformis*; *Oepikina abayi*; *Christiania taldyboensis*; *Tcherskidium* cf. *ulkuntasense*; *Prostricklandia*? sp.; *Altaethyrella otarica*; *A.* aff. *rudis*; *A. instabilis*; **Ovalospira parva*; **O. bakanasensis*; *Zygospiraella dominanta*; **Eospirigerina plana*; *Cyclospira*? *elegantula*; *Iliella minima*; *Cryptothyrella*? sp.

3. Orlovsk Horizon (middle Ashgill), Gorny Altai, Russia (Kulkov and Severgina 1989): *Glyptorthis praepulchra*; *Ptychopleurella mica*; *Austinella lebediensis*; *A.* sp. 1; *A.* sp. 2; *Plaesiomys* sp.; *Severginella altaica*; *S. schorica*; *Schizophorella fallax fallax*; *Salopina uxunaica*; *Epitomyonia* sp. 1; *Reuschella* sp.; *Triplesia mongolica*; *T. ainca*; *Oxoplecia platystrophoides*; *Anoptambonites grayae sibirica*; *Dulankarella magna*; *Diambonia septata*; *Sowerbyella* (*Sowerbyella*) *sibirica*; *Bimuria bugrychiensis*; *Strophomena* sp.; *Glyptomena subgirranensis*; *Mjoesina* cf. *rugata*; *Altaethyrella megala*; *Rhynchotrema dietkensis*; *Rostricellula exilis*; *R. sparsa asiatica*; *Salairella salairica*; *Catazyga anuensis*; *C. carteri*; *Zygospira* sp.; *Spirigerina* (*Eospirigerina*) *orloviensis*; *S.* (*E.*) *tatchalovensis*; *S.* (*E.*) *sublevis.*

4. Upper part of Goonumbla Volcanis and the upper of the Malongulli Formation (upper Eastonian–lower Bolindian, middle Ashgill), New South Wales, Australia (Percival 1978, 1979*a*, 1979*b*): *Casquella bifida*; *Paterula giganta*; *Elliptoglosa* sp.; *Doleroides* sp.; *Scaphorthis*? *aulacis*; *Leptellina* sp.; *Sowerbyella anticipata*; *Oepikina*? sp.; *Infurca tessellata*; *Kassinella anisa*; *Dulankarella*? *partita*; *Sowerbyites vesciseptus*; *Sowerbyella lepta*; *Gunningblandella resupinata.*

5. Korotkinskaya Formation (Beds 4 and 9, middle Ashgill), central Taimyr, northern Siberia, Russia (Cocks and Modzalevskaya 1997): *Ectenoglossa*? sp.; elkaniid; *Paracraniops* sp.; acrotretid; *Multispinula bondarevi*; *Hesperorthis*? sp.; *Ptychopleurella alata*; *Plectorthis* sp.; *Cyrtonotella*; *Skenidioides* sp.; *Dalmanella* sp.; *Howellites aenigmus*; *Laticrura*? sp.; *Dicoelosia* sp.; *Epitomyonia* sp.; *Ogmoplecia* aff. *plicata*; *Amphiplecia bondarevi*; *Triplesia* sp.; *Leangella* sp.; *Anoptambonites* sp.; *Eoplectodonta* aff. *rhombica*; *E.* spp.; *Sowerbyella* (*Sowerybyella*)? sp.; *S.* (*Rugosowerbyella*) sp.; *Holtedahlina* sp.; *Drummuckina*? sp.; *Geniculina* sp.; leptaenine spp.; *Fardenia*? sp.; *Parastrophina pentagonalis*; *Parastrophinella*? sp. A; *Holorhynchus giganteus*; *Tcherskidium unicum*; *T.* sp.; *Catazyga* sp.; *Cyclospira orbus*; *Plectatrypa*? *laticostata*; *Qilianotryma tajmyrica*; *Eospirigerina vetusta*; indet cyrtiinid.

6. Upper member of the Shiyanhe Formation (middle Ashgill), Xichuan, Henan Province, Central China (Xu 1996): *Orbiculoidea* sp.; **Dinorthis kassini* (including Xu's *Pionorthis* sp.); **Wangyuella* sp. (Xu's *Lordorthis* sp.); *Mimella* cf. *recta*; *Schizophorella* cf. *fallax*; *Skenidioides* sp.; *Salopina* sp.; *Howellites* sp.; **Triplesia* sp. (Xu's *Triplesia zhejiangensis*); **Aegironetes* sp. (Xu's *Leangella* (*Leangella*) sp.); *Aegiria* sp.; **Sowerbyella* (*Sowerbyella*)? sp. (Xu's *Sowerbyella* (*Sowerbyella*) *sinensis*); **Altaethyrella inflata* (Xu's *Latirhynchia inflata*); *Rostricellula xichuanensis*; *Zygospira qinghaiensis*; **Ovalospira* sp. (Xu's *Altaethyrella triplicata*); *Nalivkinia* (*Nalivkinia*) *sigangensis*; *Nalivkinia* (*Anabaria*) *xichuanensis.*

SOMERVILLE COLLEGE
LIBRARY